POCKET MECHANIC

VEHICLE MANUAL

AF816512

CITROEN RELAY
PEUGEOT BOXER
FIAT DUCATO

Diesel Models from 2001-2005

2.0 Litre HDi Engine
2.0 Litre DW10 Engine
2.0 Litre (JTD) Engine

ISBN: 9781898780984

PM3

6W6/3093

PREFACE

Small though this Workshop Manual is in size, it lacks no detail in covering most of the servicing and repair of the Citroen Relay with the 2.0-litre HDi direct injection diesel engine, the Peugeot Boxer fitted with the DW10 UTD and the Fiat Ducato fitted with the engine known as the JTD. The information in this manual is based on Citroen information.

Brief, easy-to-follow instructions are given, free from all necessary complications and repetitions, yet containing all the required technical detail and information, and many diagrams and illustrations.

Compiled and illustrated by experts, this manual provides a concise source of helpful information, all of which has been cross-checked for accuracy to the manufacturer's official service and repair procedures, but many instructions have derived from actual practice to facilitate your work. Where special tools are required, these are identified in the text if absolutely necessary and we do not hesitate to advise you if we feel that the operation cannot be properly undertaken without the use of such tools.

The readers own judgement must ultimately decide what work he or she will feel able to undertake, but there is no doubt, that with this manual to assist him, there will be many more occasions where the delay, inconvenience and the cost of having the van off the road can be avoided or minimised.

The manual is called "Pocket Mechanic" and is produced in a handy glove pocket size with the aim that it should be kept in the vehicle whilst you are travelling. Many garage mechanics themselves use these publications in their workshops and if you have the manual with you in the car you will have an invaluable source of reference.

A general faultfinding (troubleshooting) section is included at the end of the manual and all items listed are taken from actual experience, together with the necessary remedies to correct faults and malfunctioning of certain parts.

No liability can be accepted for any inaccuracies or omissions in this workshop manual, although every possible care has been taken to make it as complete and accurate as possible Every care has also been taken to prevent personal injury or damage to equipment when working on the vehicle We have tried to cover all models produced to the day of publication, but are unable to refer to all modifications and changes for certain markets or up-dating of models.

The Pocket Mechanic, PO Box 904,
Amersham, HP6 9JA, England
thepocketmechanic.co.uk

CONTENTS

0. INTRODUCTION

Our "Pocket Mechanics" are based on easy-to-follow step-by-step instructions and advice, which enables you to carry out many jobs yourself. Moreover, now you have the means to avoid these frustrating delays and inconveniences that so often result from not knowing the right approach to carry out repairs, which are often comparatively simple.

Whilst special tools are required to carry out certain operations, in this manual, we show you the essential design and construction of such equipment, whenever possible, to enable you in many cases to improvise or use alternative tools. Experience shows that it is advantageous to use only genuine parts since these give you the assurance of a first-class job. You will find that many parts are identical in the Relay and the Boxer, but our advice is to find out before purchasing new parts – **always buy your spare parts from an officially appointed dealer.**

0.0. General Information

After 13 successful years, the two Vans Citroen C25 and Peugeot J5 were replaced during February 1994 by two new models, i.e., the Citroen Relay and the Peugeot Boxer. Originally, the vans were fitted with a 1.9-litre engine or a 2.5-litre diesel or turbo diesel engine. During 2001, the 2.0-litre HDi engine was introduced. The new models are manufactured in the way of a partnership between Citroen Peugeot and also Fiat and are built in Italy.

New on the 2.0 HDi engine with "Common Rail" fuel injection is the replacement of the fuel injection pump by a high-pressure fuel pump. The new system has the following advantages.

- The ability to modulate the injection pressure between 150 and 1350 bar (2130 and 19,170 psi.).
- The ability to function at a high engine speed of up to 6000 rpm.
- Precision control of the diesel fuel injection.
- Reduction of fuel consumption.
- Reduction of pollutant emissions from the engine.

Brief Description

A separate "in-tank" low-pressure fuel delivery pump supplies the fuel to the fuel filter, from where it is directed to the high-pressure pump. Fuel under pressure is then fed to the fuel distributor rail (this is the "Common Rail"). The distributor rail stores a certain amount of fuel required by the engine, regardless of the operating range. The common rail is located between the high-pressure fuel pump and the diesel injectors and attached to the cylinder head. A high-pressure fuel regulator on the high-pressure pump regulates the pressure of the fuel at the outlet of the high-pressure pump.

The diesel fuel injectors are operated electrically by the electronic control unit (ECU) for the injection system. The amount of fuel injected depends on the following conditions.

- Duration of the electrical control by the ECU.
- Opening speed of the diesel injectors.
- Hydraulic flow of the diesel injector (number and diameter of holes).
- Fuel pressure in the fuel pressure common injection rail.

The fuel injectors are connected together by the fuel return circuit.

All models use a five-speed transmission with cable gearchange, but different transmission types are used, depending on the engine.

Various modifications were introduced since the introduction of the new models, which will be mentioned whenever necessary.

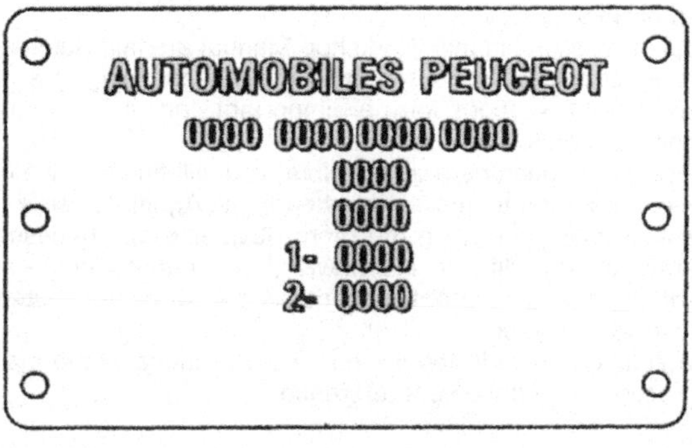

Fig. 0.1 – The type identification plate.

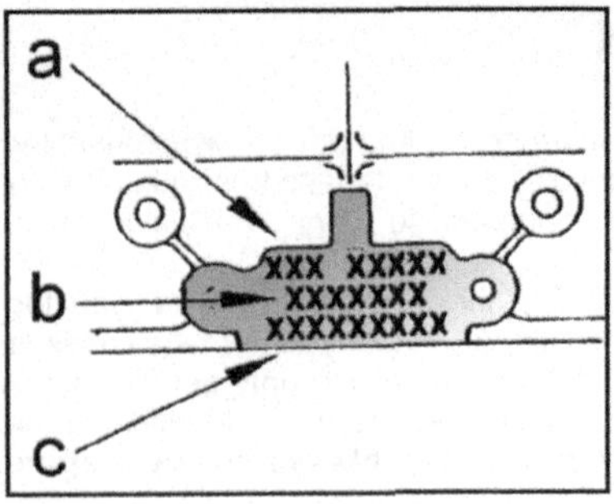

Fig. 0.2 – The location of the engine number.

0.1. Identification

The type identification plate is riveted to the centre of the upper crossmember, near the bonnet lock at the position shown in Fig. 0.1, and contains the following information: the name of the manufacturer, the chassis number, the max. permissible all-up weight and the max. weight on the front and rear axles. The chassis number can also be found in the inside of the R.H. wheel arch, immediately before the passenger seat and can be seen with the door open.

Another important identification is the engine number. The number always commences with the engine-type code. You will find the engine number always on a plate on the upper part of the cylinder block. Fig. 0.2 shows the location of the engine-type identification plate. The engine is identified with the code letters "RHV".

Important is the model year of the vehicle. This should always be quoted when parts are required.

0.2. Dimensions

The following data refer to all models irrespective of the fitted engine.

Wheelbase	
- Short wheelbase	2850 mm (113.3 in.)
- Intermediate wheelbase	3200 mm (127.14 in.)
- Long wheelbase	3700 mm (147.0 in.)
Front track	1720 mm (68.34 in.)
Rear trac	1710 mm (67.54 in.)
Overall length (typical van)	5005 mm (198.5 in.)
Overall height	2145 mm (85.22 in.)
Overall width	1998 mm (79.4 in.)
Ground clearance (depending on tyres)	170–210 mm (6.76–8.3 in.)

0.3. Filling Capacities

Fuel tank:	80 litres
Engine:	
- Without oil filter:	5.0 litres
- With oil filter:	5.3 litres
Cooling System:	
- With air conditioning system:	11.0 litres
- Without air conditioning system:	10.5 litres
Transmission:	
- ME5TU gearbox:	1.85 litres
- ME5GU gearbox:	2.75 litres
Power-assisted steering	1.3 litres

0.4. General Servicing Notes

The servicing and overhaul instructions in this Workshop Manual are laid out in an easy-to-follow step-by-step fashion, and no difficulty should be encountered, if the text and diagrams are followed carefully and methodically. The "Technical Data" sections form an important part of the repair procedures and should always be referred to during work on the vehicle.

In order that we can include as much data as possible, you will find that we do not generally repeat in the text values already given under the technical data headings. Again, to make the best use of the space available, we do not repeat at each operation the more obvious steps necessary – we feel it to be far more helpful to concentrate on the difficult or awkward procedures in greater detail. However, we summarise below a few of the more important procedures and draw your attention to various points of general interest that apply to all operations.

Always use the torque settings given in the various main sections of the manual. These are grouped together in separate sub-sections for convenient reference.

Bolts and nuts should be assembled in a clean and very lightly oiled condition, and faces and threads should always be inspected to make sure that they are free from damage burrs or scoring. DO NOT degrease bolts or nuts.

All joint washers, gaskets, tabs and lock washers, split pins and "O" rings must be replaced on assembly. Seals will, in the majority of cases, also need to be replaced, if the shaft and seal have been separated. Always lubricate the lip of the seal before assembly and take care that the seal lip is facing the correct direction.

References to the left-hand and right-hand sides are always to be taken as if the observer is at the rear of the vehicle, facing forwards, unless otherwise stated.

Always make sure that the vehicle is adequately supported, and on firm ground, before commencing any work on the underside of the car. A small jack or a make shift prop can be highly dangerous and proper axle stands are an essential requirement for your own safety.

Dirt, grease and mineral oil will rapidly destroy the seals of the hydraulic system and even the smallest amounts must be prevented from entering the system or coming into contact with the components. Use clean brake fluid or one of the proprietary cleaners to wash the hydraulic system parts. An acceptable alternative cleaner is a methylated spirit, but if this is used, it should not be allowed to remain in contact with the rubber parts for longer than necessary. It is also important that all traces of the fluid should be removed from the system before final assembly.

Always use genuine manufacturer's spares and replacements for the best results.

Since the manufacturer uses metric units when building the cars, it is recommended that these are used for all precise units. Inch conversions are given in most cases but these are not necessarily precise conversions, being rounded off for the unimportant values.

Removal and installation instructions, in this Workshop Manual, cover the steps to take away or put back the unit or part in question. Other instructions, usually headed "Servicing", will cover the dismantling and repair of the unit; once it has been stripped from the vehicle, it is pointed out that the major instructions cover a complete overhaul of all parts but, obviously, this will not always be either necessary and should not be carried out needlessly.

There are several variations in unit parts on the range of vehicles covered in this Workshop Manual. We strongly recommend that you take care to identify the precise model and the year of manufacture, before obtaining any spares or replacement parts.

Std.: To indicate sizes and limits of components as supplied by the manufacturer. Also to indicate the production tolerances of new unused parts.

O/S Paris supplied as oversize or undersize or recommended limits for such parts to enable them to be used with worn or re-machined mating parts.

U/S O/S indicates a part that is larger than Std. size. U/S may indicate a bore of a bushing or female part that is smaller than Std.

Max.: Where given against a clearance or dimension indicates the maximum allowable if in excess of the value given, it is recommended that the appropriate part is fitted.

TIR: Indicates the total indicator reading as shown by a dial indicator (dial gauge).

TDC: Top dead centre (No. 1 piston on firing stroke).

MP: Multi-purpose grease.

0.5. Jacking Up of the Vehicle

For small jobs, jack up the vehicle as described in the Owner's Manual, with the jack supplied in the vehicle. If a mobile jack is used, make sure it can be inserted between the lowest point of the vehicle and the ground. If necessary, use the vehicle jack to lift the vehicle slightly, until the mobile jack can be inserted. Due to the construction of the vehicle, a mobile jack and/or chassis stands should only be placed under the vehicle as follows, noting that a piece of rubber or wood should always be inserted between the jack head and the jacking point to protect the part in question.

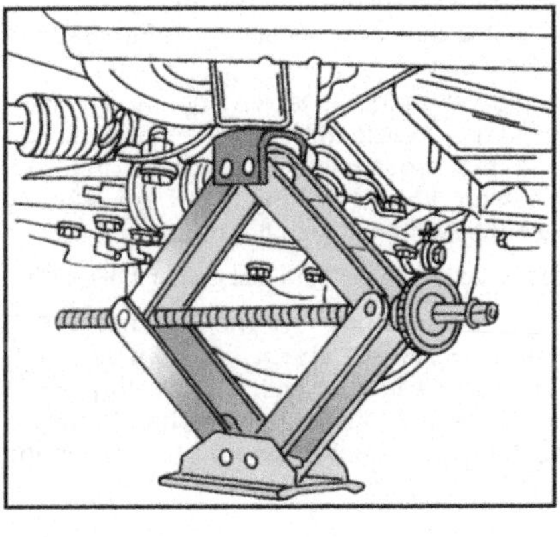

Fig. 0.3 – Lifting the front end of the vehicle with a scissor-type jack.

Never apply a mobile jack or a hydraulic jack underneath the engine oil sump, the transmission, the rear axle (except with 4DW), the front axle or underneath the front crossmember, as these parts or surrounding parts can either be damaged or distorted.

The recommendation is to first jack up one side of the vehicle, placing a chassis stand underneath the body and then jacking up the other side. The following instructions should be followed at all times.

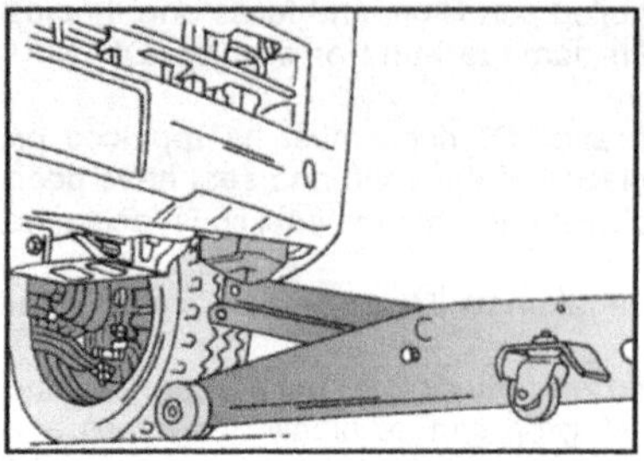

Fig. 0.4 – Lifting the front end of the vehicle with a mobile jack, as used in a workshop.

- To lift the front wheel with a scissor jack, refer to Fig. 0.3.
- To lift a front wheel with a mobile jack, use the jacking point shown in Fig. 0.3, but insert a wooden plank between jack head and jacking point. Fig. 0.4 shows this arrangement.

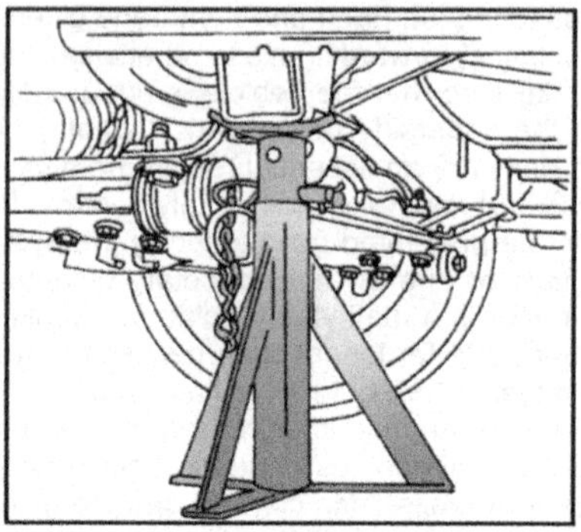

Fig. 0.5 – Chassis stands are placed in the position shown when supporting the front of the vehicle.

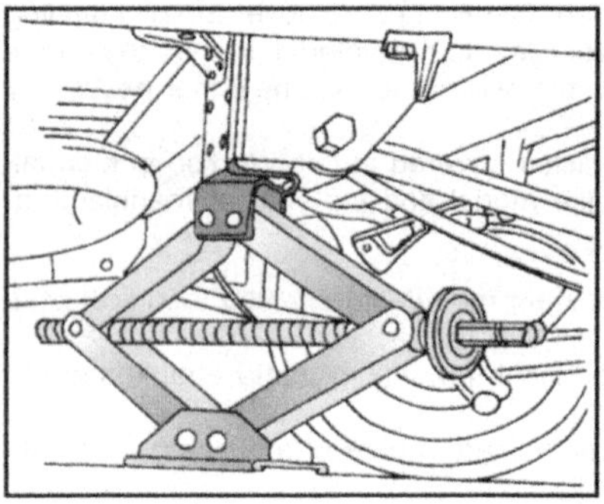

Fig. 0.6 – Jacking up the rear end of the vehicle with a scissor jack.

- Chassis stands are placed underneath the front end of the vehicle at the same positions as the jack. In this case, the lifting jack must be placed underneath the inner crossmember. This will enable you to insert the chassis stands. A properly placed stand is shown in Fig. 0.5. Note that chassis stands with flat head are used. If you only have stands with "V"-shaped head, fill the "V" in a suitable manner. Make sure the vehicle cannot slip off the stands.

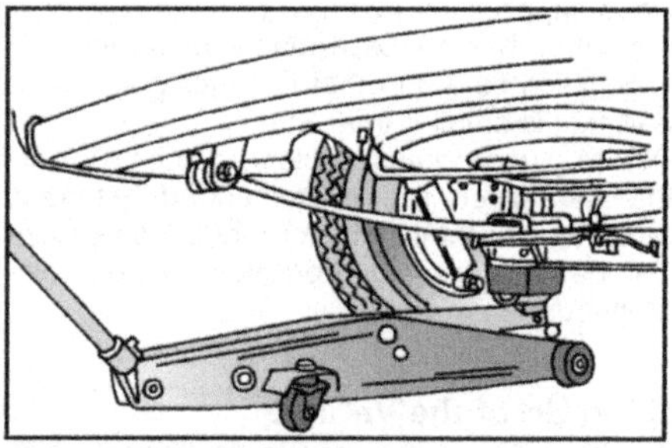

Fig. 0.7 – Jacking up the rear end of the vehicle with a mobile jack as used in the workshop.

The rear end of the vehicle is jacked up and supported in the following manner.

- A scissor jack is used as shown in Fig. 0.6. The jacking point is located before the rear spring.
- A mobile jack is arranged as shown in Fig. 0.7. The jacking point is the same as shown in Fig. 0.6, but a wooden plank should be placed between jack head and jacking point.

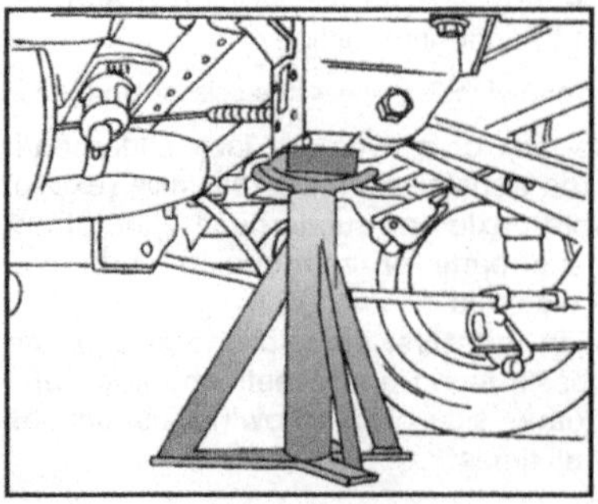

- Chassis stands are placed under the jacking points. To insert the chassis stands, lift the vehicle at the connection between the rear spring and the crossmember, to place the jack underneath the position shown in Fig. 0.8. As already explained, use only chassis stands with flat heads.

Fig. 0.8 – Chassis stands placed underneath the rear end of the vehicle.

Make sure that the chassis stands and the jack are strong enough to carry the vehicle. Always check the front and rear wheels on the other side of the vehicle with a brick, to prevent the vehicle from rolling off the jack. Also, apply the handbrake.

Fig. 0.9 shows two further examples where chassis stands can be placed underneath the vehicle.

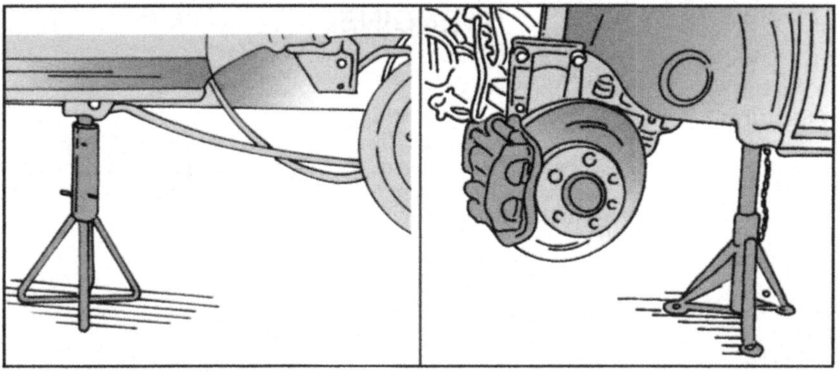

Fig. 0.9 – Placing chassis stands underneath the vehicle. On the L.H. side underneath the rear spring bracket, and on the R.H. side at the front of the vehicle.

Note: *It is always difficult to raise a vehicle first on one side and then on the other. Take care that the vehicle cannot tip-over when the first side is lifted. Ask a helper* to *support the vehicle from the other side.*

1. ENGINE

1.0. Main Features

Type:	Four-cylinder diesel engine with direct injection (high-pressure diesel injection), transversely mounted, overhead valves (2 per cylinder) and camshaft in cylinder head, driven by toothed belt. Cylinder head is made of aluminium alloy.
Number of cylinders:	4
Engine capacity:	1997 c.c.
Cylinder bore:	85.00 mm (3.3371 in.)
Piston stroke:	88.00 mm (3.4962 in.)
Compression ratio:	17.6:1
Max. Power:	66 kW (90 BHP) at 4000 rpm
Max. Torque:	19.2 kgm (138.2 ft.lb.) at 1900 rpm
Oil pressure	See the "Lubrication" section
Fuel injection type	See relevant section
Valve Timing	
Inlet valve opens	9° BTDC
Inlet valve closes	23° ABDC
Exhaust valve opens	40° 6′ BBDC
Exhaust valve closes	6° 6′ ATDC

BTDC = Before dead centre
ATDC = After top dead centre
BBDC = Before bottom dead centre
ATDC = After top dead centre

1.1. Engine – Removal and Installation

The engine and transmission are removed from the vehicle as a complete unit. In the workshop, a special lifting table is used to take out the unit, which will not be described in the following text. The power unit is removed towards the front, i.e., the front end of the vehicle must be removed as necessary to lift out the assembly. You will need a hoist or small crane, which must have the necessary carrying capacity. The same applies to the lifting brackets. Fig. 1.1 shows a suitable appliance to lift out the unit.

Hose clamps are requiring a special pair of pliers to open and tighten them. Otherwise, cut them and use screw-type clamps during installation.

Removal can now be attempted as follows.

- Disconnect the battery cables. If a radio with anti-theft code is fitted, make sure you have the code handy for later. Also, remember that pre-programmed stations will be lost. The battery can be removed if preferred.
- Drain the cooling system (engine cold). A drain plug is inserted into the bottom of the radiator. Open the expansion tank cap to speed-up the draining. The anti-freeze can be collected if it appears to be in good condition. A plug can be fitted into the cylinder block. In this case, remove it.

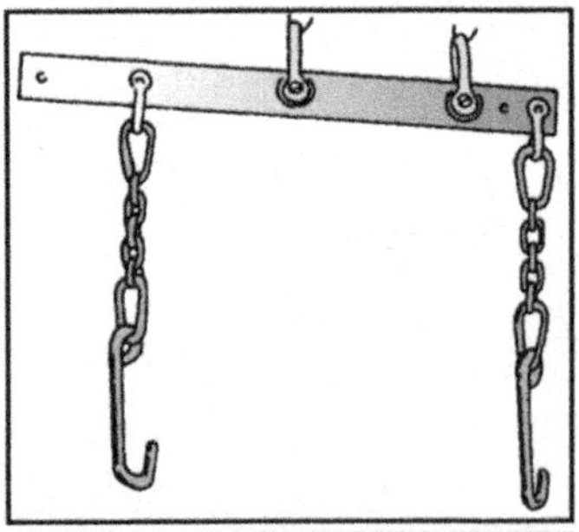

Fig. 1.1 – A lifting bracket and suitable chains can be used to lift the engine and transmission out of the engine compartment.

- The bonnet should be removed or placed in a vertical position and tied up.
- Drain the transmission oil.
- Unscrew the four bolts shown by the arrows in Fig. 1.2 and remove the complete front of the vehicle.

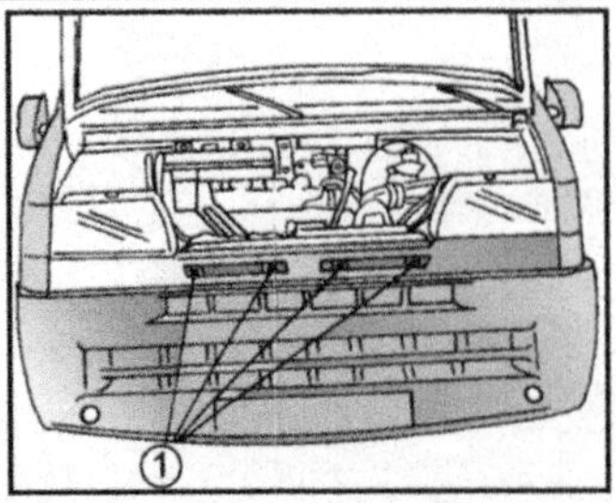

Fig. 1.2 – Remove the screws (1) and take off the front panel.

- In the inside of the wheel arches, remove the protective panel shown in

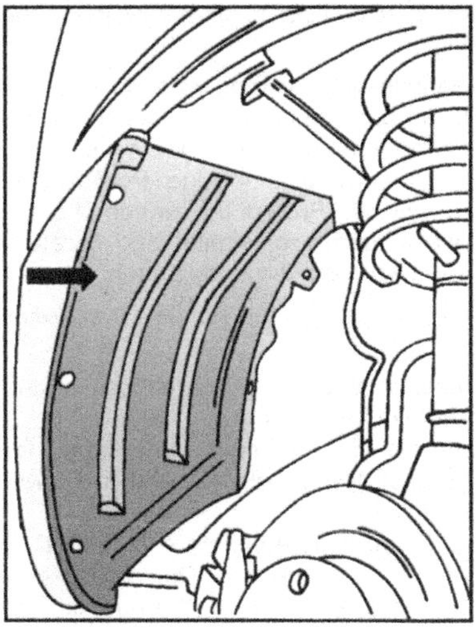

Fig. 1.3 – Remove the protective panels in the inside of each wheel arch. The support strut for the front bumper, shown in Fig. 1.10, can also be removed.

Fig. 1.3. The panel (mud shields) are secured by plastic pegs, i.e., the workshop uses a pair of special pliers to remove the pegs. You will see a stay between the body and the front bumper, which must be unscrewed at the top and the bottom (see also Fig. 1.10). The bumper is now removed by referring to Fig. 1.5.

- Fig. 1.4 shows a view of the engine compartment. The following operations are carried out by referring to the illustration. Note the differences between models with and without air conditioning systems.

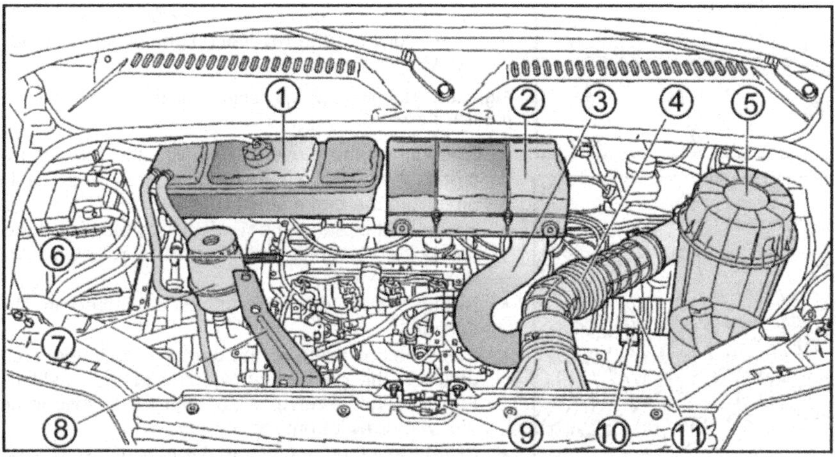

Fig. 1.4 – Details for the removal of the engine. The numbers are referred to the text.

- Fig. 1.4 First, disconnect the connector (10). Then, remove the individual parts in the specified order: the air cleaner (5), the air flow meter (11) and the air inlet duct (4). Next, remove the air duct (3) and the trim cover (2).
- Remove the support (8) with the steering fluid reservoir tank (7). Remove the coolant reservoir tank (1). The bonnet locking device (9) must also be removed.

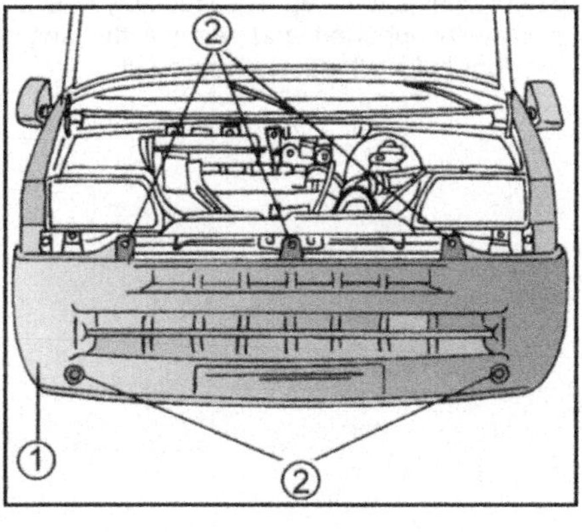

- Disconnect the upper and lower hoses of the cooling radiator and the flexible hose (6).
- Disconnect all electrical harnesses and, depending on the model, the headlamp connectors.
- In the case of vehicles without air conditioning system, uncouple the cooling fan connectors.

Fig. 1.5 – The front bumper (1) is secured with screws (2) at the positions shown.

- Remove the direction indicators. Remove the following parts, some of which are shown in Fig. 1.6: the screws (1) and the front panel, the screw (3), the attachment of the electrical harnesses on the cooling fans, the radiator, the attachment of the hydraulic control unit for ABS (depending on model), the screws (2) and the lower crossmember.

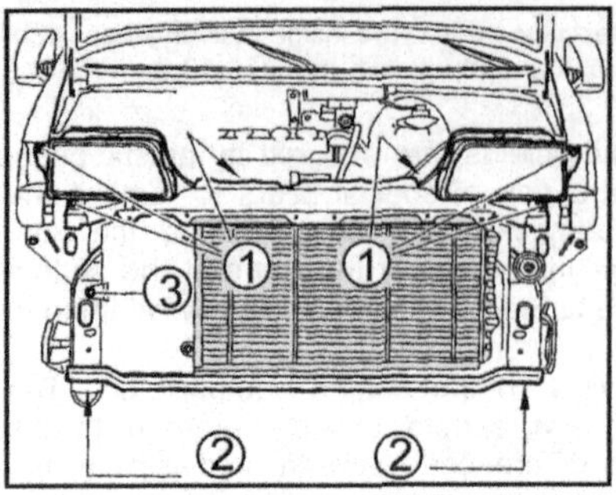

Fig. 1.6 – Models without air conditioning system. Remove the numbered items as described.

- On models with air conditioning system, refer to Fig. 1.7 and remove the parts as applicable. First, disconnect the cooling fan connectors and remove the direction indicators.
- Remove the screws (1) and take out the front panel. Remove the screws (3) and (4). Protect the air conditioning condenser with a piece of carton or similar and carefully move the condenser (5) away from the body panel.

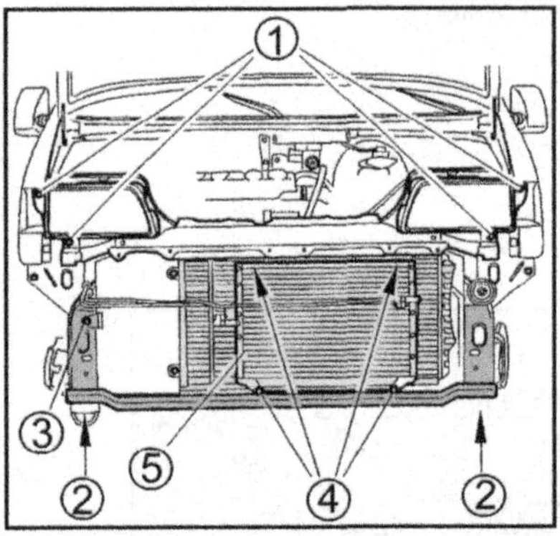

Fig. 1.7 – Models with air conditioning system. Remove the individual items as described.

- Remove the attachments of all electrical harnesses on the cooling fan and then remove the radiator.
- Remove the attachment of the hydraulic control unit for ABS (depending on model), the screws (2) and the lower crossmember.
- On all models, remove the auxiliary equipment drive belt.
- Remove the air conditioning compressor. The compressor must be suspended on a piece of wire on the L.H. side of the vehicle.
- Disconnect the cables from the starter motor. Pull off the connector from the reversing light switch. If fitted, withdraw the ABS connectors.

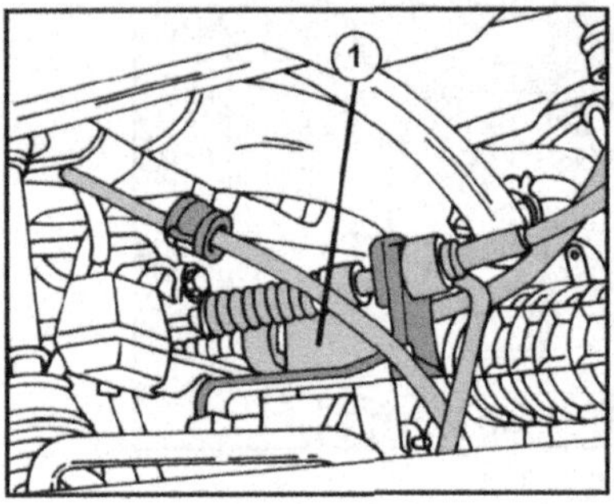

Fig. 1.8 – The gear change cable support bracket (1) is located at the position shown.

- Disconnect and move to one side the fuel feed and return pipes, the clutch slave cylinder, the earth cable and the pipe from the brake servo unit.
- Remove the attachment of the connecting tube between the power steering pump and the rotary valve and then remove the tube.
- Remove both drive shafts as described in the relevant section.
- Push the ball joints off the gearchange levers (screwdriver), unscrew the gearchange control at the position shown in Fig. 1.8 and remove. Also, undo the speedometer cable and withdraw it from the transmission.
- Disconnect the cables from the rear of the alternator.
- Disconnect the inlet and outlet hoses from the heater radiator.
- Disconnect the cable from the oil pressure switch.
- Unscrew the torque strut at the rear of the engine and separate the exhaust pipe from the exhaust manifold.
- In the workshop, the power unit is now lifted with the lifting table referred to above and removed towards the front. Under DIY conditions, you will need ropes or chains to lift up the assembly until it is no longer under tension. Make sure that the power unit is well-supported and remove the engine mountings by referring to Fig. 1.9. The three bolts (1) and the four bolts (2) must be removed.

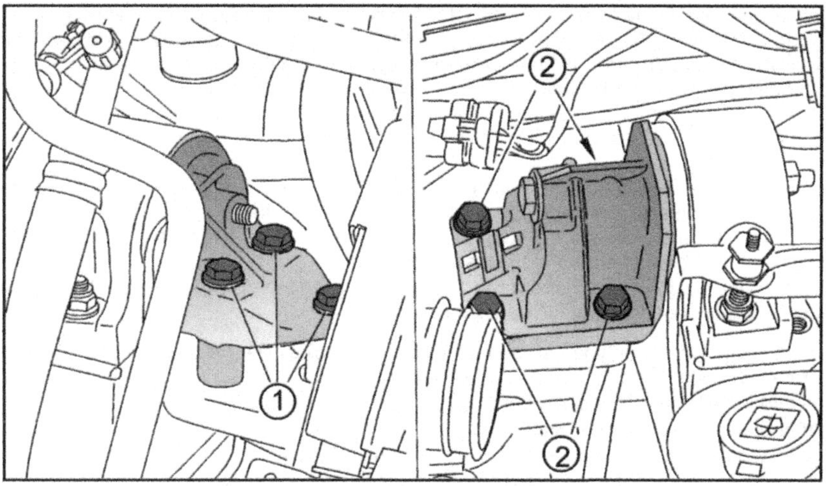

Fig. 1.9 – The engine mountings. Bolts (1) and (2) must be removed.

- Slowly lift the assembly towards the front out of the engine compartment. Check once more that all connections, hoses, etc., have been freed before lifting out the unit.

The installation of the engine and transmission is a reversal of the removal procedure. The following points should be noted in particular.

- Always replace self-locking nuts, lock plates, lock washers and other parts.
- Prepare the engine compartment to lift the engine and transmission in position without difficulties. Loose cables, hoses, etc., should be pushed well aside and fastened with sticky tape or tank tape.
- Lift the engine and transmission in position and tighten the engine mountings. These are shown in Fig. 1.9. Tighten the bolts and nuts with the torque values given in the illustrations. The bolts (2) are tightened to 8.0 kgm (58 ft.lb.) after the mounting is in the correct position. The bolts (1) are tightened to 5.0 kgm (36 ft.lb.).

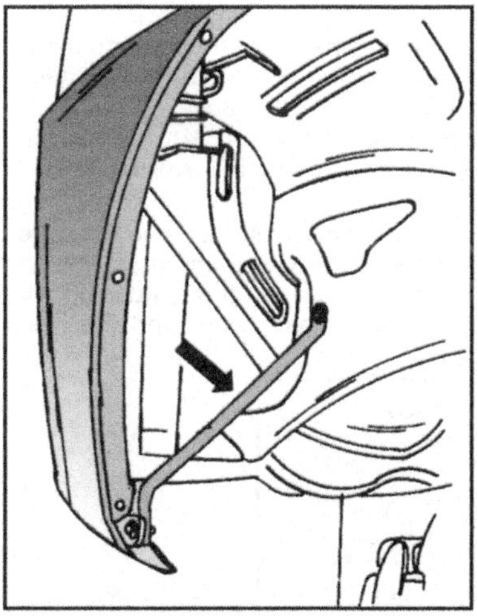

Fig. 1.10 – In each wheel arch, you will find a stay for the front bumper, which must be refitted as shown before fitting the protective panels.

- Refit the reaction rod. The gearbox end is tightened to 9.0 kgm (65 ft.lb.) and the engine subframe end to 5.0 kgm (26 ft.lb.).
- Refit the drive shafts as described in Chapter 4.
- Refill the transmission, the steering fluid container and the cooling system as described in the relevant section. The cooling system must be bled of air.
- After fitting the power unit, start the engine and move the steering wheel from one lock in the other, to bleed the system. Check the fluid level in the container and correct it, if necessary. Lower the vehicle onto its wheels after the steering is free of air. Tighten the wheel bolts.

1.2. Engine – Dismantling

Before commencing the dismantling of the engine, all exterior surfaces should be cleaned, as far as possible, to remove dirt or grease. Plug the engine openings with clean cloth first to prevent any foreign matter entering the cavities and openings. Detailed information on engine dismantling and assembly is given in the sections dealing with servicing and overhaul (sections commencing at 1.4.), and these should be followed for each of the sub-assemblies or units to be dealt with.

Follow the general dismantling instructions given below.

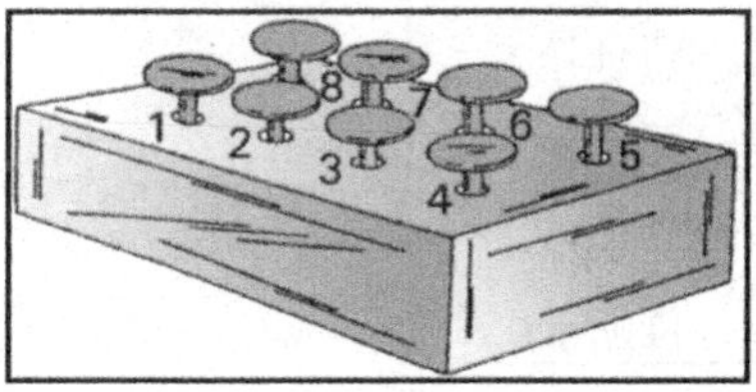

Fig. 1.11 – Valves can be pushed through the bottom of an upside-down cardboard box.

- Dismantling must be carried out in an orderly fashion to ensure that parts, such as valves, pistons, bearing caps, shells and tappets, are replaced in the same positions as they occupied originally. Mark them clearly but take care not to scratch or stamp on any rotating or bearing surfaces. A good way to keep the valves in order is by piercing them through an upside-down cardboard box, as shown in Fig. 1.10 and writing the number against each valve (refer to previous page). Segregate together the tappets, the springs and retainers with collets for each valve, if possible in small plastic bags for each individual valve.
- If a proper engine dismantling stand is not available, it will be useful to make up wooden support blocks to allow access to both the top and bottom faces of the engine. The cylinder head, once removed from the block, should be supported by a metal strap, screwed to the manifold face and secured by two nuts onto the manifold studs.
- The normal order of removal of parts for a complete engine strip-down is given below but this may, of course, be modified if only partial dismantling is required.

A complete strip-down of the engine can be carried out as follows. The following operations are described under separate headings.

- Removal and installation of the drive belt.
- Removal and installation of the cylinder head.
- Removal and installation of the camshaft and the camshaft oil seal.
- Removal and installation of the toothed belt.
- Removal and installation of the oil pump.
- Thoroughly clean the exterior of the engine to remove all loose dirt and oil. Brush away all dirt from the joint faces and take off the transmission unit, if still fitted.

Remove the parts shown in Fig. 1.12 from the exhaust side of the engine.

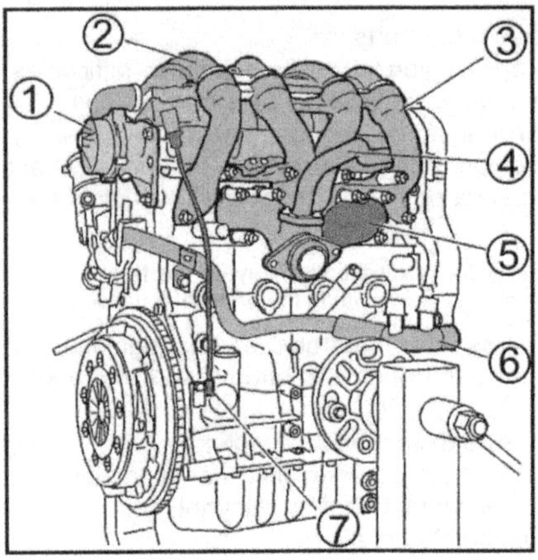

Fig. 1.12 – Removal of parts on the exhaust manifold side of the engine.

- Detach the exhaust gas recycling tube (4) from the exhaust manifold and the airbox (2). Remove the TDC sensor (7), the vacuum pump (1) and the airbox (2). Remove the exhaust manifold (5), the inlet manifold (3) and the exhaust gas recycling pipe (4) at the same time. The coolant inlet housing (6) and the connecting hoses are also removed.
- On the other side of the engine, remove the parts shown in Fig. 1.13. These are the fast idle cable (1), the diesel fuel filter and its housing (2), the cover (3) and the thermostat, the coolant outlet housing (4), the injection high-pressure supply lines (5) and the ventilation hose (6).
- Remove the crankshaft pulley bolt and remove the pulley from the end of the crankshaft. Also from the front of the engine, remove the high-pressure pump housing, the upper timing belt cover, the lower timing belt cover, the upper cylinder head cover and the lower cylinder head cover.

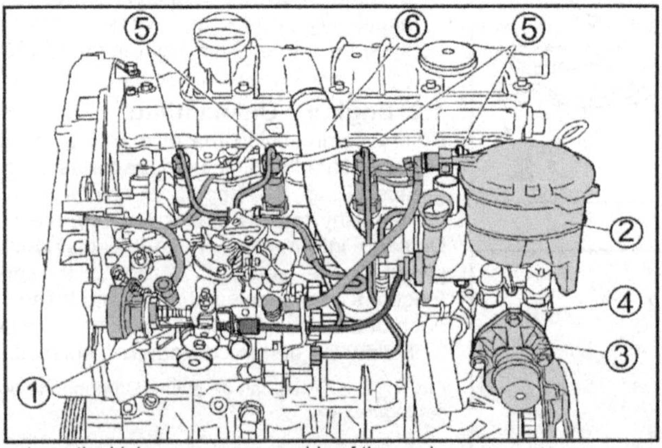

Fig. 1.13 – Removal of parts on the high-pressure pump side of the engine.

- Mark the position of the clutch plate in relation to the flywheel by punching two marks in the clutch and the flywheel on opposite points. Unscrew the clutch and remove the driven plate. The flywheel should be held against rotation by inserting a screwdriver into the flywheel teeth or by means of a toothed sector as shown in Fig. 1.14. If the driven plate looks in good condition, take care not to contaminate it with oily fingers. The flywheel can now be unscrewed.

Fig. 1.14 – Locking a flywheel.

- Remove the timing belt as described under the relevant heading.
- Remove the oil filter, the bolt in the centre of the oil cooler, the oil cooler and the oil pressure switch. Fig. 1.15 shows the location of the parts.

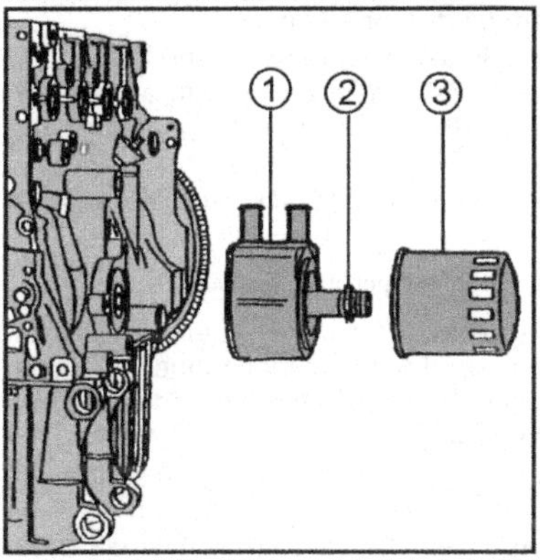

Fig. 1.15 – The oil filter (2) with the oil cooler (1). Remove the bolt (2).

- Remove the cylinder head as described under the relevant heading.
- The next operation is the removal of the high-pressure injection pump. Again, a separate section deals with this operation.
- Remove the oil sump. A steel sump is fitted with a seal, and an aluminium sump is fitted without seal.
- Remove the front cover from the engine as described during the replacement of the front crankshaft oil seal.
- Remove the oil pump (see the "Lubrication System" section).
- Turn the cylinder block upside down, with the open crankcase uppermost and rotate the crankshaft until two of the crankpins are in the bottom dead centre position.
- Mark the two accessible bearing caps and the connecting rods with a centre punch and remove the connecting rod bearing cap nuts. Remove the bearing caps and bearing shells and put the parts aside. Use a hammer handle and push the pistons out of the cylinder bores. The upper bearing shell will be stuck either to the connecting rod or to the crankpin. Retrieve the shell and immediately fit the other shell and the bearing cap to the removed connecting rod/piston assembly.
- Rotate the crankshaft until the remaining crankpins are at the bottom dead centre and remove the pistons and connecting rods in the same manner.

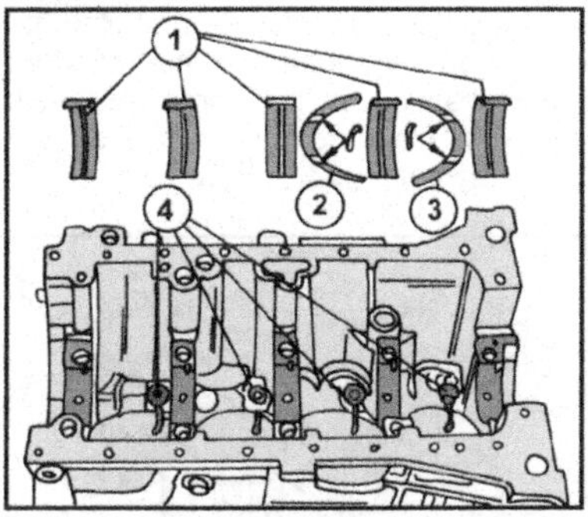

Fig. 1.16 – The crankshaft bearings.

1 Main bearing shells
2 Thrust half washer
3 Thrust half washer
4 Oil splash jets

- Mark the main bearing caps and their bearing locations on one side of the engine and remove the caps one after the other. Remove the bearing shells out of the caps, thoroughly clean them and mark the back of each cap with the bearing number.
- Carefully lift the crankshaft out of the bearing bores and remove the five bearing shells from the crankcase. Also, remove the two half thrust washers from the second crankshaft main bearing journal (see Fig. 1.16). Attach the bearing shells of each bearing to the respective main bearing cap, if the same shells are to be fitted. Also, mark where each half of the thrust washer has been removed from.
- Dismantle the pistons as described in Section 1.4.1.1. and the cylinder head as described in Section 1.4.0.2. Note that oil splash jets are fitted to the cylinder block. Fig. 1.16 shows where they are located. The tightening torque for the jets is 1.0 kgm (7 ft.lb.).

1.3. Engine – Assembly

When assembling the engine, follow the general proceedings outlined in this section and refer to the later sections, commencing at 1.4. for any detailed information that is necessary. In the following description, it is assumed that all sub-assemblies have been overhauled or replaced, if necessary.

Carefully clean the cylinder block, paying particular attention to the crevasses, which are easily overlooked. Note that the cylinder block is made of aluminium alloy and should, therefore, not be scraped in order to clean the gasket faces.

Lubricate each rotating or moving part with engine oil BEFORE it is assembled. Lubricating it after the assembly has little purpose, as the oil cannot reach the gliding surfaces at all times. Oil seals should be replaced as a matter of course. The same goes for the lock washers or lock plates. Check that all oil ways are thoroughly clean.

Assemble the engine in the following general order.

- Assemble the pistons and connecting rods as described in Section 1.4.1.1.
- Fit the bearing shells into the connecting rods and the big end bearing caps, if the original parts are re-used. Make sure the lugs on the shells engage into the holes in the bearing bore.
- Generously lubricate the main bearing shells with oil and place them into the crankcase bores. These are the shells with the oil grooves.
- Fit the oil splash jets to the bottom of the cylinder block at the positions shown in Fig.1.16.
- Fit the half thrust washers to the No. 2 main bearing (No. 1 is next to the flywheel) so that the washer friction faces (with the oil grooves) are against the crankshaft flange. Generously lubricate the thrust washers with oil.
- Once more, check that the bearing shells are well-lubricated and slowly lift the crankshaft into the crankcase. Rotate the crankshaft a few times to settle the bearing shells. Fit the lower bearing shells into the bearing caps (half thrust washers to No. 2 bearing cap) and fit the bearing caps in accordance with the marks made on dismantling. The numbering must be on the side of the oil filter. The rear cap is not fitted at this stage.

The crankshaft end float is now checked and, if necessary, corrected, as described below.

- Fit a dial gauge with a suitable bracket to the front end of the cylinder block and set the stylus and place the stylus onto the end of the crankshaft, as shown in Fig. 1.17. If a dial gauge with a magnetic case is used, place the base against the crankshaft end face, with the stylus resting against the cylinder block face.
- Move the crankshaft to one side as shown by the arrow in the illustration and set the dial gauge to "Zero".

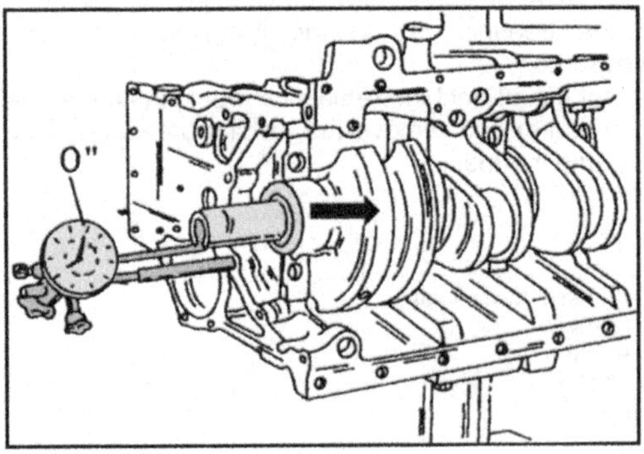

Fig. 1.17 – Checking the crankshaft end float.

- Move the crankshaft to the other side in the same manner and read the dial gauge indication. The value should be between 0.07 and 0.32 mm (0.0003–0.013 in.).
- If the end float exceeds the upper value, replace the half thrust washers. The crankshaft must be partially removed to take out the washers. Washers are available in thicknesses of 2.35, 2.40, 2.45 and 2.50 mm, apart from the standard thickness. Two washers of the same thickness must always be fitted to one side of the crankshaft.

A special tool is required to fit the rear main bearing cap to prevent damage to the cap side seals. First, coat the mounting faces for the bearing cap (on the crankcase face) with sealing compound. Then, proceed as follows.

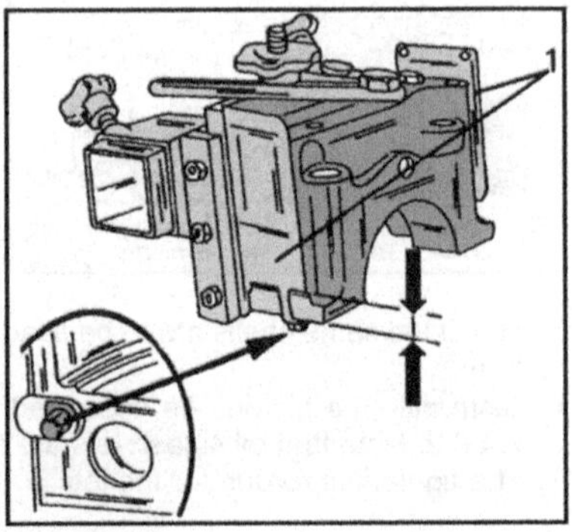

Fig. 1.18 – Fitting the side seals of the rear bearing cap. The guide plates (1) push the seals into the grooves. The seals must protrude at the bottom as shown in the inset.

- Fit the two guide plates to the special tool, as shown in Fig. 1.18. The screws are only tightened finger-tight.
- Fit two new side seals to the sides of the rear bearing cap with the holder. Fit the holder to the main bearing cap, using the clamp screw. Lubricate the guide plates with engine oil.

- Push the guide plates together with one hand and insert the whole assembly at an angle into the cylinder block opening, until the cap is in the correct position. Push the tool downwards until this is the case.
- Fit the two main bearing cap bolts and tighten them finger-tight.
- Unscrew the tool body from the cap and remove the tool together with the guide plates from the cylinder block.
- Commencing at the centre and working towards the outside, evenly tighten all main bearing cap bolts to 7.0 kgm (50.5 ft.lb.). First, tighten the bolts of the centre bearing cap, then the bolts of No. 4 cap, followed by the bolts of No. 2 cap, then the bolts of the front cap and finally the bolts of the bearing cap on the flywheel side.
- Fit two of the flywheel bolts into the crankshaft flange and insert a screwdriver between the bolts to rotate the crankshaft. Check the crankshaft for binding before proceeding with the assembly.
- Coat the outside of a new rear oil seal with oil and drive it in position into the crankcase and the rear bearing cap. Take care not to damage the oil seal. Use a mandrel that will cover the complete outer face of the seal.

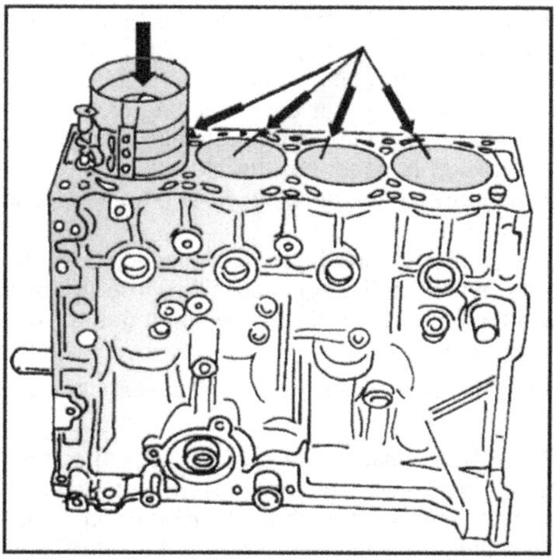

Fig. 1.19 – Fitting a piston.

- Arrange the assembled piston and connecting rods in their cylinder order. Fit a piston rind clamp band around the piston rings, as shown in Fig. 1.19, and guide the connecting rod into the cylinder bore and over the crankpin (bearing shell must be in connecting rod).
- Check that the piston bowl is in the correct position and use a hammer handle to push the piston into the bore. The clamp band will slide off the piston as soon as all piston rings have entered the bore. Make sure that the connecting rod is guided over the crankpin at the same time. The cylinder block can be placed on its side to facilitate the installation. Two of the crankshaft pins should be at the bottom dead centre position, and connecting rod/piston assemblies of these two should be fitted.
- Fit the pistons of the other cylinders in the same manner.
- Fit the bearing shells in accordance with the marks made during dismantling into the bearing caps of the two fitted connecting rods and lubricate the shell faces with oil. Fit the bearing caps over the connecting rod bolts and check that both cylinder marks on rod and cap are on the same side of the cylinder block.
- Fit new nuts and evenly tighten the two nuts to 2.0 kgm (14.5 ft.lb.). From the final position, each nut must then be tightened by a further 70°. Either estimate the angle (1/4 of a turn is 90°) or use a graduated disc. Rotate the crankshaft a few times to check for binding and then rotate the shaft until the two remaining crankpins are at the bottom dead centre.
- Fit the other two pistons and connecting rod assemblies in the manner described, always checking that the marking is strictly adhered to.

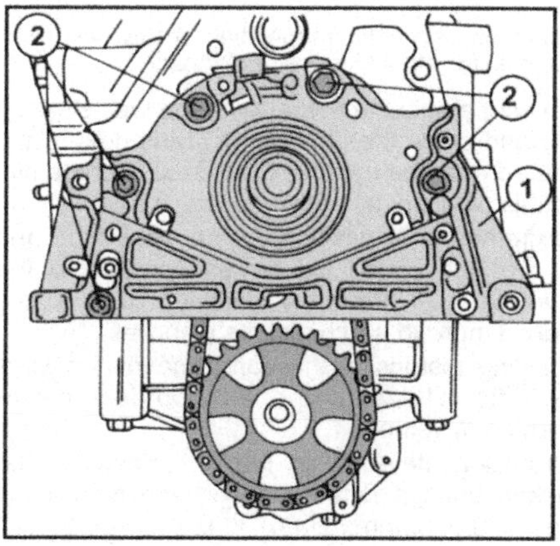

- Fit the oil pump drive sprocket with the key and the drive chain to the end of the crankshaft. The cylinder block should be with the cylinder head gasket face uppermost, with the chain hanging down, engaged with the sprocket.

Fig. 1.20 – The front oil seal cover (1) is secured in position with 6 screws (2). Tighten to 1.2 kgm.

- Fit the front oil seal cover. The sealing face must be coated with a sealing compound. Tighten all screws to 1.2 kgm (9.5 ft.lb.). Fig. 1.20 shows the fitted cover with the location of the screws.
- When fitting the oil pump, engage the chain over the oil pump sprocket, place the oil pump in position and slide the spacer between the pump and the cylinder block.

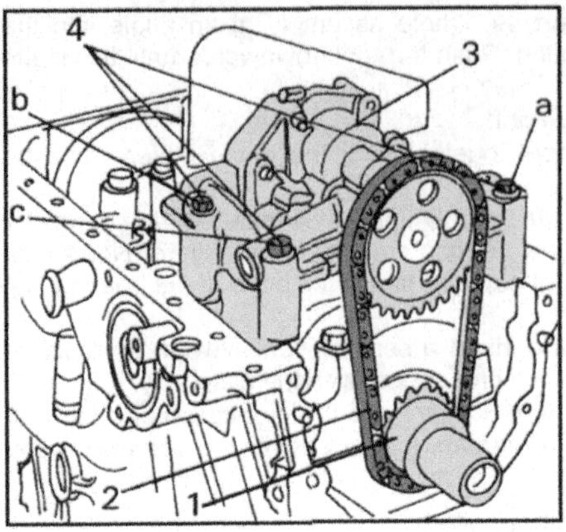

Fig. 1.21 – The location of the various securing screws for the oil pump.
1 Crankshaft sprocket
2 Drive chain
3 Oil pump
4 Securing screws
a Bolt, 80 mm long
b Bolt, 70 mm long
c Bolt, 65 mm long

Fit the oil suction unit. The oil pump bolts are tightened to 1.8 kgm (14 ft.lb.). Fig. 1.21 shows the pump in a fitted position. Note that the length of bolts (a), (b) and (c) is not the same.

- Coat the outside and the sealing lip of the front crankshaft oil seal with engine oil and fit the seal over the end of the crankshaft and into the oil seal cover. Use a suitable mandrel, which covers the outside face of the seal, as shown in Fig. 1.22 on the next page. Take care not to damage the seal.
- Fit the oil sump with a new gasket and tighten the 23 screws evenly to a torque setting of 1.6 kgm (11.5 ft.lb.). Insert the screws in accordance with the markings during removal. An aluminium oil sump is guided on the crankcase by means of a dowel pin. No gasket is used, but the block face must be coated with sealing compound. Again, insert the screws as marked during removal. The same torque setting applies.
- Fit the flywheel. Always use new bolts. Rotate the flywheel until all bolt holes are aligned. Coat the threads with "Loctite" before installation. Tighten the bolts evenly to 5.0 kgm (36 ft.lb.). The flywheel must be held against rotation. Either insert a screwdriver into the teeth of the flywheel ring gear or use a locking sector as shown in Fig. 1.15.

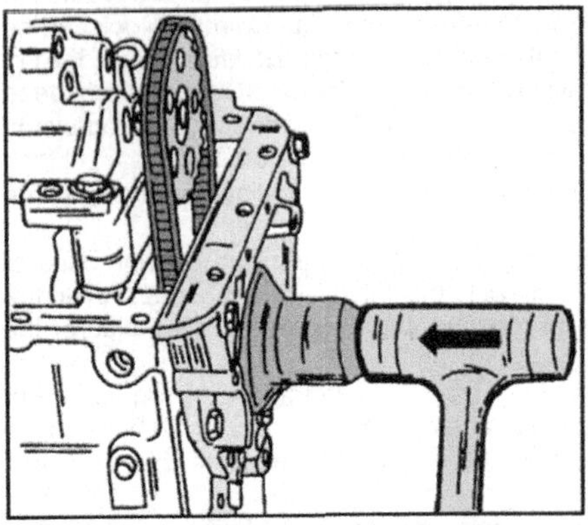

Fig. 1.22 – Fitting the front crankshaft oil seal after installation of the oil pump.

- Fit the clutch-driven plate and the clutch pressure plate. Align the clutch-driven plate with a centring mandrel (mandrel sets are available from tool hire companies and one will fit the driven plate). Tighten the clutch plate securing screws to 2.0 kgm (14.5 ft.lb.).

Cylinder head gaskets are available in different thicknesses and the gasket to be used depends on the protrusion of the piston above the cylinder block face. To check the protrusion, rotate the crankshaft until each piston in turn is at top dead centre position. Measure the amount the piston end is above the cylinder block face (protrusion) as follows.

Fig. 1.23 – Measuring the protrusion of the piston above the cylinder block face. The stylus is placed onto the block face and set to "Zero".

- Place a dial gauge with a suitable holder onto the well-cleaned cylinder block face and place the dial gauge stylus next to the piston to be measured onto the cylinder block face. Set the dial gauge to "Zero" in this position, as shown in Fig. 1.23.
- Leave the dial gauge holder in position on the block face and swing the stylus over the piston, as far as possible to the centre, as shown in Fig. 1.24. Read the indicated value on the dial gauge and make a note to which pistons it applies.
- Measure the remaining pistons in the same manner, always rotating the crankshaft until each piston is at top dead centre position. Write down each value.
- The greatest dimension determines the thickness of the cylinder head gasket, but the difference between two of the pistons must not exceed 0.12 mm (0.0047 in.).

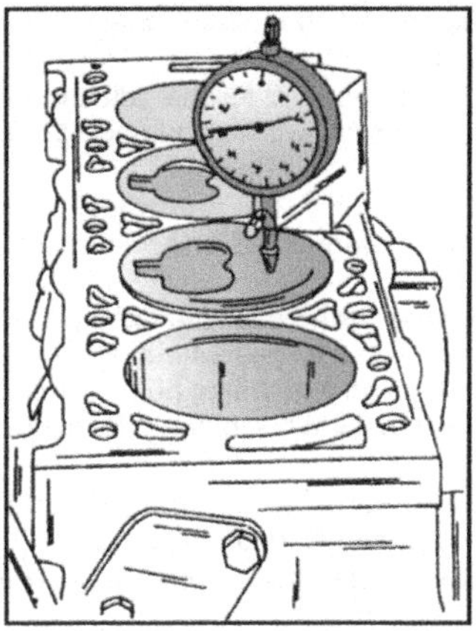

The protrusion must now be evaluated with the help of Figs. 1.23 and 1.24 to select the correct gasket. The 5 different gaskets are available and their thickness must be selected by referring to Fig. 1.25. Important is the number of holes at position "A" and at position "B".

Fig. 1.24 – Measuring the protrusion of the pistons above the cylinder block face. The stylus is placed on the piston crown, as far as possible near the centre of the piston.

- If the greatest dimension is between 0.51 and 0.55 mm, fit a gasket that has two holes at position "A" and one hole at position "B". The thickness of the gasket is 1.26 mm (+/−0.04 mm).
- If the greatest dimension is between 0.55 and 0.59 mm, use a gasket with 2 holes at position "A" and two holes at position "B". The thickness of the gasket is 1.30 mm (tolerances as above).
- If the greatest dimension is between 0.59 and 0.63 mm, use a gasket with 2 holes at position "A" and three holes at position "B". The thickness of the gasket is 1.34 mm (tolerances as above).

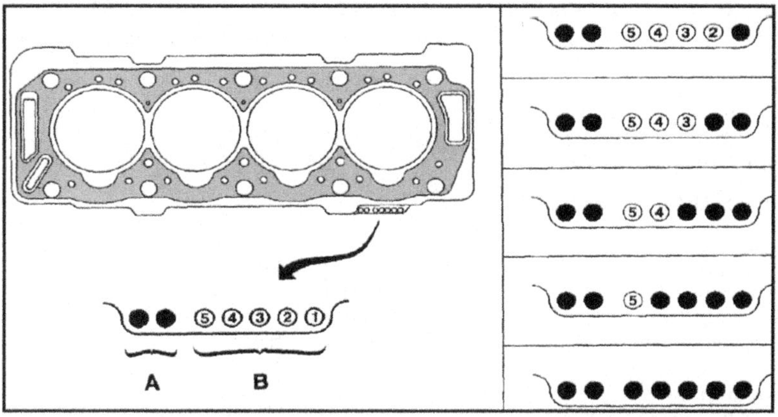

Fig. 1.25 – View of the cylinder head gasket. Count the number of holes at positions "A" and "B" to select the correct gaskets. Also, measure the thickness of the gasket.

- If the greatest dimension is between 0.63 and 0.67 mm, use a gasket with 2 holes at position "A" and four holes at position "B". The thickness of the gasket is 1.38 mm (tolerances as above).
- If the greatest dimension is between 0.67 and 0.71 mm, use a gasket with 2 holes at position "A" and five holes at position "B". The thickness of the gasket is 1.42 mm (tolerances as above).

Note: As can be seen from the above instructions, you will have to follow them to the letter in order to select the correct gasket. We recommend, therefore, to measure the protrusion as described and go to the parts department to have the correct gasket selected. Remember there are five thicknesses, and mistakes can be made.

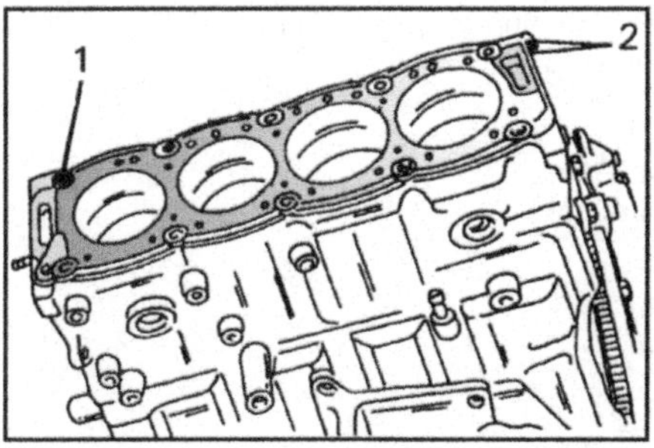

After the protrusion has been measured, turn the crankshaft until all four pistons are approx. in the centres of their bores, before continuing with the assembly.

Fig. 1.26 – View of the fitted cylinder head gasket, with the position of the dowel sleeve (1) and the identification notches (2).

- Check that the dowel pin (1) in Fig. 1.26 is in position in the cylinder block face and place the cylinder head gasket in position. The notches on the outer edge of the cylinder head gasket must be located on the flywheel side of the engine. No sealing compound must be used for the gasket. Fit it dry.
- Fit the cylinder head in accordance with the instructions in Section 1.4.0.4. and check and, if necessary, adjust the valve clearances as described in Section 1.4.0.5.
- Fit the water pump with a new gasket and tighten the screws to 1.5 kgm (11.5 ft.lb.).
- Fit the thermo switches into the adapter housing for the thermostat and tighten them to 1.5 kgm (11.5 ft.lb.).

- Fit the thermostat with two new gaskets and attach the adapter housing to the thermostat housing. Fit the complete thermostat housing with a new gasket to the cylinder head.
- Remove the camshaft timing gear (which should be still in position, as it is required for the adjustment of the valve clearance) and fit the camshaft oil seals with a suitable mandrel, without damaging the seal lips. Coat the seal lips with a little grease.
- Fit the camshaft timing gearwheel. Tighten the centre bolt to 4.5 kgm (32.5 ft.lb.). The gearwheel must be held against rotation. To do this, insert a metal rod into one of the timing gear holes and rest the rod against the head.
- Fit the timing belt. First, place the belt over the crankshaft timing gearwheel, then over the fixed pulley in the centre, the injection pump drive gear and finally over the camshaft timing gearwheel. Hold the driving side of the belt tight and then place it over the tensioning roller and the water pump wheel. Check that the timing belt has engaged correctly with all gearwheels and tension the belt as described in the relevant section.
- Fit the crankshaft pulley. Coat the bolt thread with "Loctite". The bolt is tightened to 4.0 kgm (30 ft.lb.). From this position, tighten the bolt a further 60°. The crankshaft must be prevented from rotation.
- Fit the two timing belt covers with the spring clips. The clip in the centre is pushed in position with a large screwdriver.
- Refit the remaining parts to the engine. Tighten the exhaust manifold nuts to 2.0 kgm (14.5 ft.lb.) and the inlet manifold nuts to 2.3 kgm (17 ft.lb.). Fit the oil filter with a new sealing ring and tighten it to 1.4 kgm (10 ft.lb.).

1.4. Engine – Servicing and Overhaul

1.4.0. CYLINDER HEAD AND VALVES

1.4.0.0. Technical Data

Cylinder Head:	
- Cylinder head height, face to face	133 mm +/– 0.05 mm
- Max. distortion of head surface	0.03 mm
- Grinding of head surface	Not permissible
- Valve faces below cylinder head face	See under "Valves"
Cylinder Head Bolt Length	
Length between underside head and bolt end	133.5 mm
Bolts longer than 133.5 mm	Replace
Camshaft Bearing Bores in Cylinder Head:	
- No. 1, flywheel end	27.500–27.533 mm
- No. 2	28.000–28.033 mm
- No. 3	28.500–28.533 mm
Camshaft Bearing Bores in Cylinder Head, Repair Size:	
- No. 1, flywheel end	28.000 (+0.033 mm)
- No. 2	28.500 (+0.033 mm)
- No. 3	29.000 (+0.033 mm)
Cylinder Head Gasket	See description during installation of the cylinder head.
Valve Guides	
Nominal outside diameter	14.020–14.030 mm
Repair size 1 – 1 groove	14.320–14.43 mm
Locating Bore in Cylinder Head:	
- With nominal outside diameter	13.981 mm
- Repair size 1	14.281 mm
Inside diameter of guides	8.020 mm
Installation height, guide to head face	36.5 +/– 0.5 mm
Valve Shaft Clearance in Guides:	
- Inlet valves	0.015–0.230 mm
- Exhaust valves	0.035–0.250 mm

Valve Seats

Seat angle, inlet and exhaust valves	90°
Width of valve seat faces – Inlet/Exhaust	2.45/3.00 mm
Valve Seat Outer Diameters – Inlet Valves:	
- Nominal diameter	39.837 mm
- Repair size 1	40.337 mm
Valve Seat Outer Diameters – Exhaust Valves:	
- Nominal diameter	33.820 mm
- Repair size 1	34.300 mm
Locating Bore in Cylinder Head – Inlet valves:	
- Nominal diameter	39.700 mm
- Repair size 1	40.200 mm
Locating Bore in Cylinder Head – Exhaust Valves:	
- Nominal diameter	33.700 mm
- Repair size 1	34.200 mm

Valves

Valve Shaft Diameters:	
- Inlet valves	7.985 mm
- Exhaust valves	7.985 mm
Valve seat angle	90°
Valve Head Diameters:	
- Inlet valves	38.50 +/– 0.20 mm
- Exhaust valves	33.00 +/– 0.20 mm
Min. Valve Length:	
- Inlet valves	112.4 mm
- Exhaust valves	111.85 mm
Stem end, max. grinding allowance	0.20 mm
Stem to guide clearance, max.	0.10 mm
Valve Shaft Clearance in Guides:	
- Inlet valves	0.015–0.230 mm
- Exhaust valves	0.035–0.250 mm

Valve Springs

Wire diameter	3.70 mm

Valve Timing

Inlet valve opens	9° BTDC
Inlet valve closes	23° ABDC
Exhaust valve opens	40° 6′ BBDC
Exhaust valve closes	6° 6′

Camshaft

Camshaft end float	0.025–0.114 mm
Camshaft bearing running clearance	0.04–0.082 mm
Camshaft Journal Diameter (standard):	
- No. 1	27.50 mm
- No. 2	28.00 mm
- No. 3	28.50 mm
Camshaft Journal Diameter (Repair Diameters):	
- No. 1	28.00 mm
- No. 2	28.50 mm
- No. 3	29.00 mm
Camshaft identification	Colour ring on No. 1 cam

1.4.0.1. Cylinder Head – Removal

The cylinder head can be removed with the engine fitted to the vehicle, but it should be pointed out that the timing belt must be partially removed and dial gauge with a magnetic base must be available to determine the thickness of the cylinder head gasket required, when the gasket is replaced. Provided that these conditions are met, proceed as follows. Fig. 1.27 shows the engine compartment with the location of some of the parts to be disconnected or removed.

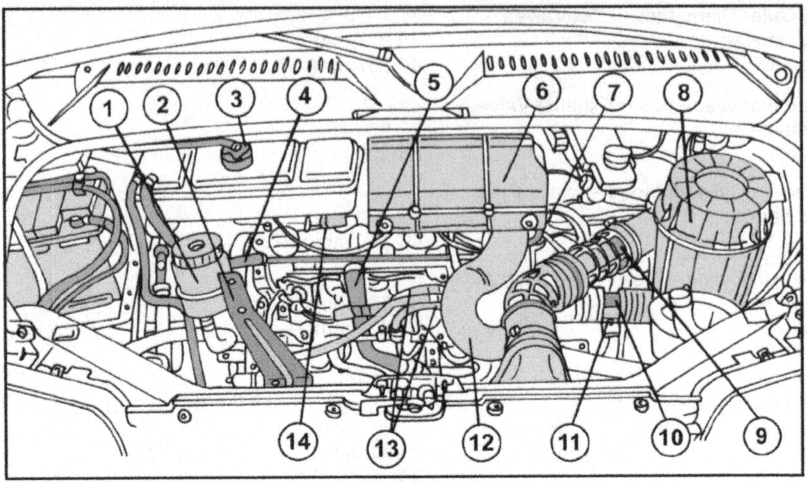

Fig. 1.27 – View of the engine compartment. The numbers are referred to the description.

- Disconnect the battery earth cable and drain the cooling system.
- Place the front end of the vehicle on chassis stands and remove the protective panel underneath the vehicle.
- Disconnect the connector plug (11) from the air flow meter.
- Remove the air cleaner (8), the air flow meter (10) and the air inlet duct (9) in the order given. Also, disconnect and remove the air duct (12).
- Slacken the air inlet collar at location (7) on the throttle valve housing.
- Remove the engine trim cover (6).
- Remove the expansion tank (3) for the cooling system.
- Disconnect the following parts and push them to one side: the ventilation hose (4), the two fuel hoses (13) and the hose (5) for the oil vapour re-circulation.
- Unscrew the securing bracket (2) and remove the oil reservoir for the steering fluid.
- Remove the EGR valve assembly with the heat exchanger (14).
- Disconnect all electrical leads, wires and all cooling hoses from the cylinder head. The coolant outlet housing must be removed.
- Remove the R.H. wheel.
- Remove the following items in accordance with the description under separate headings: drive belt at the front of the engine, toothed timing belt, high-pressure injection rail (common rail) and the turbo charger.
- Attach the engine with a rope or chain to a suitable hand crane or similar lifting device and lift the power unit out of its mountings until the R.H. engine mounting can be removed. Only remove the two upper bolts next to the timing wheel.
- Remove the cylinder head cover.

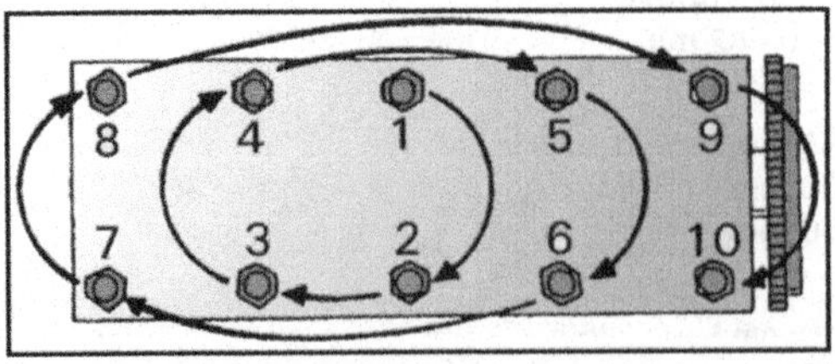

Fig. 1.28 – The cylinder head bolts are tightened in the order shown. Reverse the order when the bolts are slackened.

- Slacken the cylinder head bolts in the reverse order of Fig. 1.28 in several stages and remove the bolts. ***Note that the length of each bolt must be measured before re-use.*** As the cylinder head is located by means of a dowel pin, it may be difficult to remove the head. Workshops use two angled levers, as shown in Fig. 1.29, which are inserted into the side of the head. Lift the two levers together until the head is free of the cylinder head gasket. Lift off the head completely and remove the cylinder head gasket. Also, remove the dowel pin if it is attached to the cylinder head. It must be tapped into the cylinder block before the head is refitted.

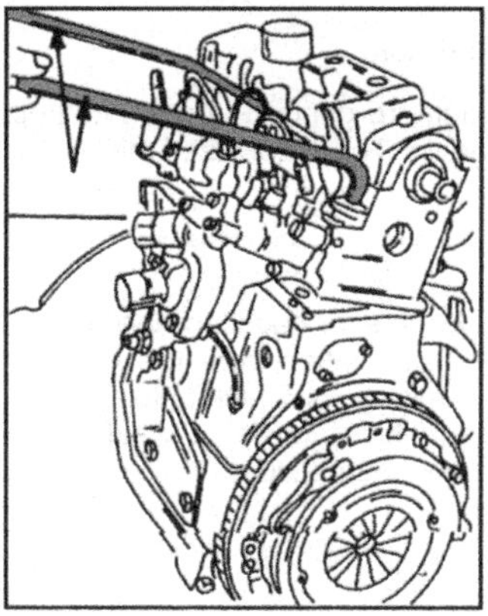

Fig. 1.29 – Removal of the cylinder head with the two special levers.

- Clean the cylinder block and cylinder head surfaces with a suitable solvent. Never use a scraper or emery cloth to clean the gasket faces. Re-grinding the cylinder head face is not permitted. If the cylinder head must be dismantled, refer to the next section.

1.4.0.2. Cylinder Head – Dismantling

Fig. 1.30 shows the major parts of the cylinder head. If the head is replaced, the cylinder head must be completely dismantled.

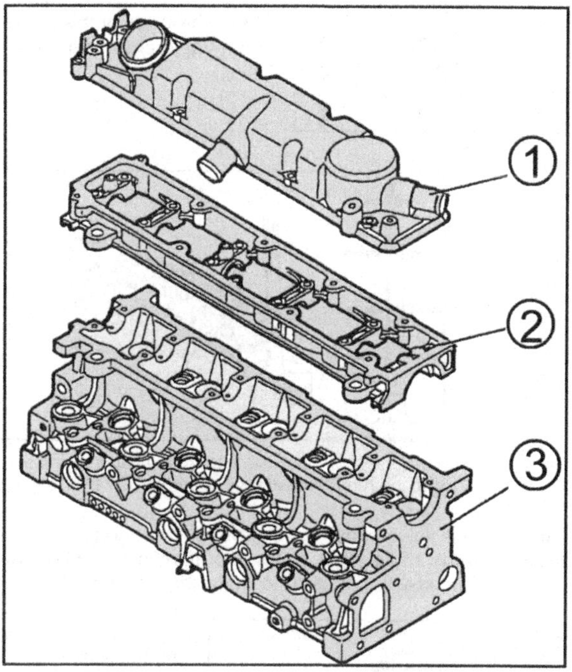

Fig. 1.30 – The three major parts of the cylinder head.

1 Cylinder head cover
2 Camshaft housing
3 Cylinder head

- Remove the camshaft housing as described later on and lift out together with the camshaft. This is carried out as shown in Fig. 1.31. Remove the camshaft oil seal. The seal must always be replaced even if it appears in good condition.
- Remove the valve tappets. Mark the fitted location of each tappet to refit them into their original bores.

Fig. 1.31 – The camshaft housing (1) and the fitted camshaft are removed as an assembly.

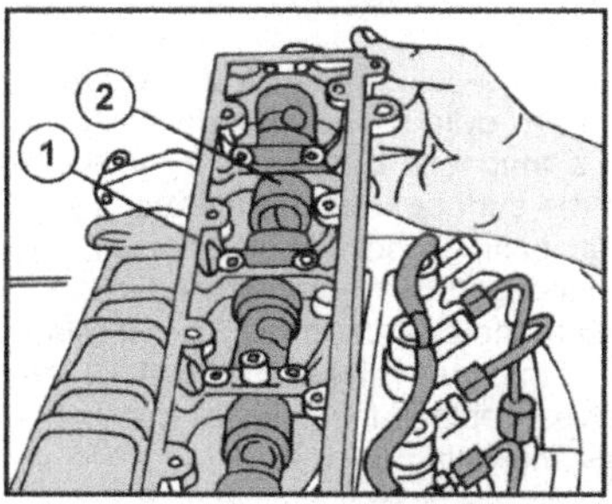

A valve lifter is required to remove the valves from the cylinder head. Valve lifters are available from most tool hire companies and there should be no problem obtaining one. Fig. 1.32 shows a suitable valve lifter, which can be used. It is also possible by placing a piece of wood into the combustion chamber, to support the valve head and then to proceed as follows.

- Use a piece of tube, the same diameter as the valve spring cap, and place the tube over the valve spring. With a hammer, give a short blow to the tube. The valve springs will be compressed and the valve cotter halves remain inside the tube. The operation may be slightly tricky, due to the inclination of the valves. Take the parts off the valve and withdraw the valve from the other side. If fitted, remove the valve stem seals. Mark the valves with the number. No. 1 valve is the nearest to the timing end of the engine. All parts belonging to one valve should be kept together.

Fig. 1.32 – Using a valve lifter.

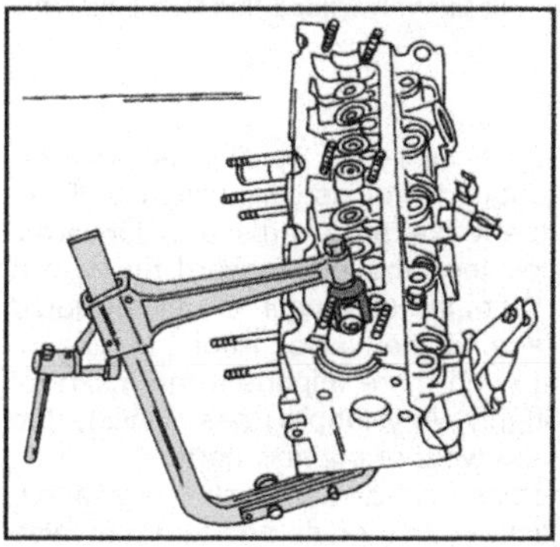

1.4.0.3. Cylinder Head Overhaul

Refer to Section 1.4.0.0. and compare the values given for the engine in question.

Valves: Check the valve faces for wear or grooving. If the wear is only slight, re-grind the valves to their appropriate angle in a valve grinding machine. All valves have the same angle. The valve head thickness, shown in Fig. 1.33, must not be less than 0.5 mm.

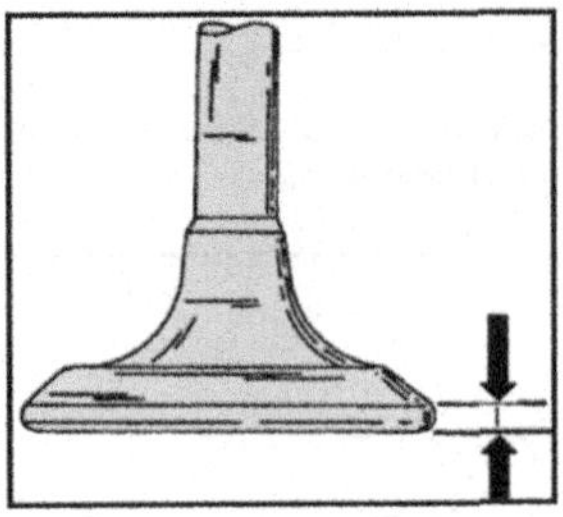

Fig. 1.33 – Observe the minimum permissible value of 0.5 mm, measured between the arrows.

Check the valve stem diameters and in this connection the inside diameters of the valve guides. If there is a deviation from the nominal values, it may be necessary to replace the valve guides (see below). Also, check the ends of the valve stems. There should be no visible wear in this area. Valve ends can be re-grounded to a maximum value.

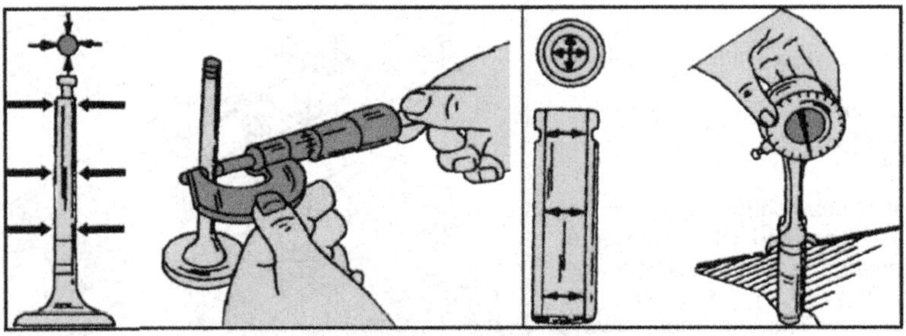

Fig. 1.34 – By measuring the outer diameter of the valve stems and the inner diameter of the valve guide bores, you will be able to establish the running clearance of the valve stems in the guide. You will need the necessary micrometres.

Valve Seats: Valve seats can be re-machined to their original angle with the appropriate cutters. All seats have the same angle (45°). If this operation is carried out properly, there should be no need to grind-in the valves. Use correction cutters to bring the valve seating area in the centre of the valve head. Make sure that the valve seat, given in Section 1.4.0.0., is obtained. We recommend, however, to have the valves re-furbished in an engine shop.

After valve seats have been re-cut, measure the distance between the valve heads and the cylinder head surface. To do this, place a dial gauge with a magnetic base onto the cylinder head face, with the stylus on the surface, and set the dial gauge to "Zero". Move the magnetic base to set the stylus onto the valve head surface, as shown in Fig. 1.35. The indicated value is the height from the valve head to the cylinder head face. If this is not 0.5–1.00 mm in the case of the inlet valves or 0.90–1.40 mm in the case of the exhaust valves, replace the cylinder head. The measurement is carried out as shown in Fig. 1.35.

Fig. 1.35 – Measuring the height of the valve head surfaces below the cylinder head face.

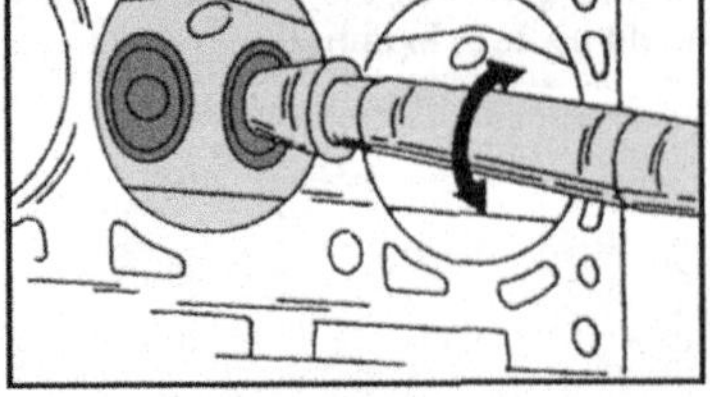

Fig. 1.36 – Grinding-in of valves.

Valves should be ground into re-cut seats. To do this, coat the valve seat with lapping compound and use a suction tool, as shown in Fig. 1.36. Move the valve backwards and forwards. Every so often, lift the suction tool, move it forward by ¼ of a turn, and continue grinding. Work the seat until an uninterrupted ring is visible around the face of the valve. After grinding-in, clean the cylinder head, and even more important the inside of the valve guide bores thoroughly. Any lapping paste inside the cylinder head will accelerate the wear of the new parts.

Use a pencil and mark across the valve seat closely spaced. Drop the valve into the respective valve guide and turn the valve by 90°, using the suction tool, applying slight pressure to the tool. Remove the valve and the pencil marks have been removed from the entire circumference. The gap created will indicate the width of the valve seat and can be measured with a ruler or caliper. Otherwise, repeat the grinding until this is the case.

Valve Springs: Springs should be compared with new springs and rejected if weak or cracked. Lay each spring on a surface plate and check that the ends of the springs are square. Slide the originally fitted spring and a new spring together over a long bolt (with large washers under bolt head and nut) and screw a nut onto the end of the bolt. Clamp the bolt head into a vice and slowly tighten the nut. If the coils of the original spring close before the coils of the new spring, replace the springs as a set, as the other springs will have similar defects.

Valve Guides: Valve guides and valve stems should first be inspected for visible wear. Clean the inside of the valve guides with a rag moistened in petrol, inserting the rag into the guides and move it to and fro to clean the inside of the bore. Valve stems are best cleaned by clamping them into an electric drill and using a wire brush to clean off any carbon deposits.

Measure the inside diameter of the guides. As an inside micrometre is necessary for this operation, which is not always available, you can insert the valve into its guide and withdraw it until the valve head is approx. level with the cylinder head face. Rock the valve to and fro and check the play. Although no exact values are available, it can be assumed that the play should not exceed 1.0–1.2 mm (0.04–0.047 in.).

Valve guides are available with oversize outside diameters (see Section 1.4.0.0.), and the locating bores in the cylinder head must be bored out to take the new guides. This may be a job for an engine shop. Drive out the guide with a suitable mandrel. The cylinder head can be heated in boiling water to facilitate the operation. Measure the protruding end of the guide (write it down) before driving the old guide out and drive in the new guide to obtain the same dimension.

New valve guide always requires the fitting of new valves, and the valve seats must be re-cut and the valves ground into the seats.

Valve guides must be reamed after installation to the diameter given in Section 1.4.0.0. Use an adjustable reamer. The correct running clearance of the valve stem is automatically obtained when the guides have the correct internal diameter.

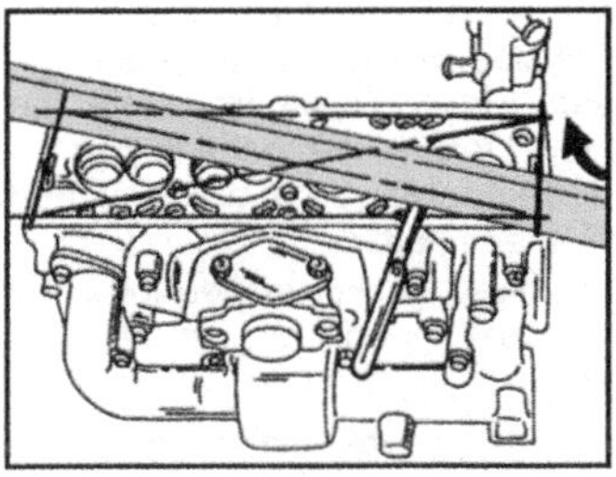

Fig. 1.37 – Checking the cylinder head gasket face for distortion. Check in the directions shown.

Cylinder Head: Thoroughly clean the cylinder head surface and measure the surface for distortion. To do this, place a fine-edge steel ruler across the cylinder head face as shown in Fig. 1.37 and measure the gap between ruler and head with feeler gauges. A feeler gauge of more than specified in Section 1.4.0.0. can be inserted. If the gap is greater, replace the cylinder head. Consult a dealer, however, before replacing a cylinder head.

Camshaft: A camshaft is allocated to a certain engine. Check the bearing journal diameters with a micrometre. If the measurements are outside the limits, fit a new camshaft.

Place the camshaft with the two end journals into V-blocks or clamp the shaft between the points of a lathe and apply a dial gauge to the centre journal. Slowly rotate the camshaft and read off the dial gauge. If the indication is more than 0.01 mm, replace the shaft, as it is bent.

To check the camshaft end float, place the shaft into the cylinder head without fitting the tappets in order to avoid the pressure of the valve springs onto the camshaft.

Arrange a dial gauge, with the stylus of the gauge resting against the end of the camshaft. Move the camshaft to and fro and read off the dial gauge. The dial gauge reading must not exceed the value given in Section 1.4.0.0. for the end float.

1.4.0.4. Cylinder Head – Assembly

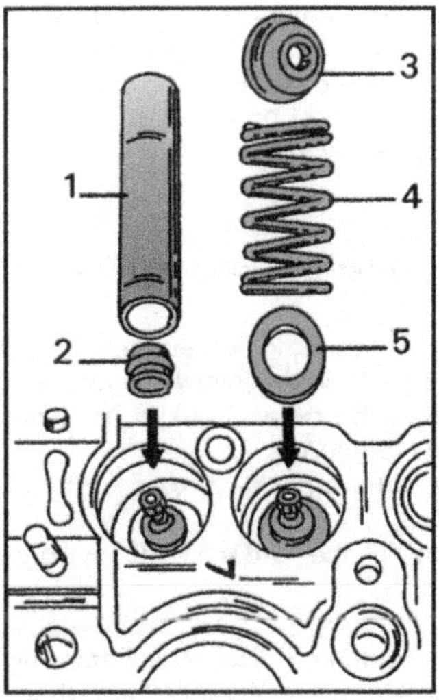

Fig. 1.38 – Fitting a valve.
1 Tube to fit valve stem seals
2 Valve stem seal
3 Valve spring cup
4 Valve spring
5 Washer

- If the cylinder head has been replaced, transfer all parts from the old to the new head. Thoroughly clean all parts and lubricate all moving parts with engine oil.
- Insert the valves into their guides. If valves have been replaced, fit them into the respective guides. The same applies to the originally fitted valves.
- Assemble the valves. First, place the valve spring seat over the cylinder head and push the valve stem seals over the valve stems and the valve guides, as shown in Fig. 1.38. Fit the upper valve spring cup over the valve spring and compress the valve spring with a valve spring compressor. When the valve stem end protrudes from the valve spring cup, insert the two valve cotter halves with a pair of pointed pliers. Slowly release the valve spring compressor, checking continuously that the valve cotter halves are kept in position.
- With a plastic mallet tap against the end of all valve stems, if the valve cotter halves have not engaged properly, they will jump out. A piece of rag should be placed over the valve stem ends – just in case.
- Lubricate the valve tappets with engine oil and insert them into the original bores (in accordance with their markings).
- Coat the bearing journals and the cams with engine oil and insert the camshaft into the bearing bores. The camshaft housing is fitted as described later on.
- Refit all other parts to the cylinder head in reverse order to the removal.

1.4.0.5. Cylinder Head – Installation

Thoroughly clean the threaded bores for the cylinder head bolts. The bores must be free of grease and dry. If necessary, use a threaded tap to re-cut the thread. The threads must be cleaned after re-cutting them.

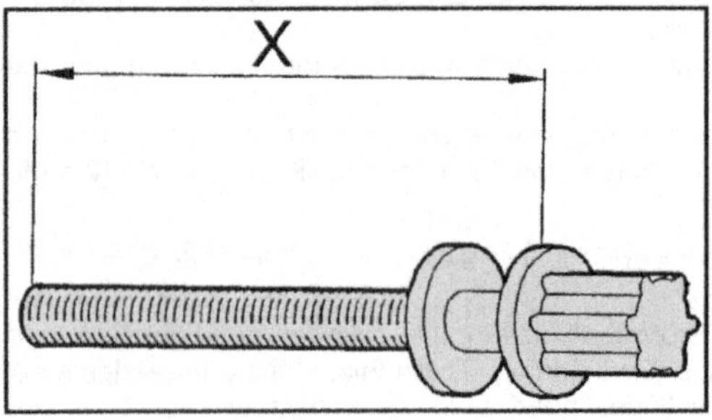

- Rotate the crankshaft until all pistons are in the centre of their stroke. You can check the position of the key in the crankshaft timing gear, which must be in a horizontal position.

Fig. 1.39 – The length of the cylinder head bolts is measured between the points shown (X).

- Check the length of each cylinder head bolt between the underside of the bolt head and the end of the thread. Fig. 1.39 shows where the measurement should take place. If the length "X" is more than 133.3 mm, the bolts are stretched and new bolts must be fitted.

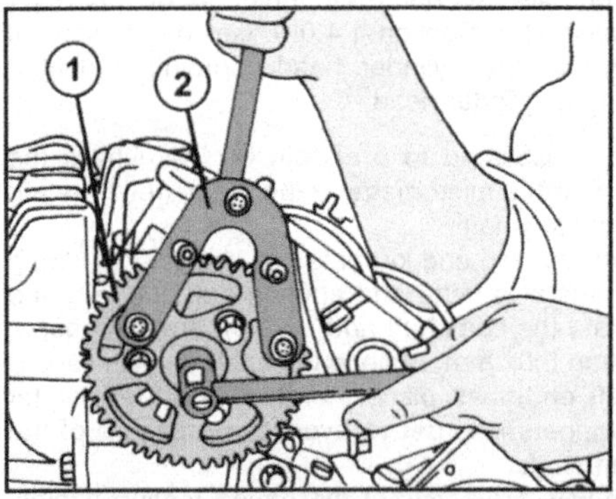

- Place the new cylinder head gasket in position. The thickness identification must be on the high-pressure pump side of the engine.

Fig. 1.40 – The camshaft timing gear can be held as shown when the centre bolt is slackened or tightened. A similar lever can be attached.

- Prevent the camshaft gearwheel (pulley) from rotating. A simple lever, as shown in Fig. 1.40, can be used. Undo the gearwheel bolt and remove the camshaft gearwheel. The gearwheel must be removed in order to adjust the camshaft position sensor.

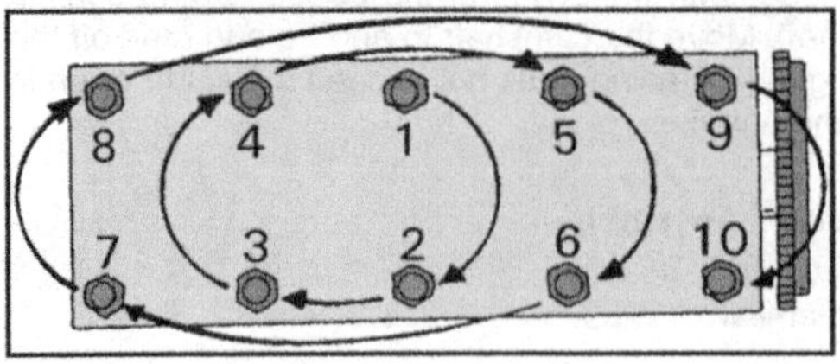

Fig. 1.41 – Tighten the cylinder head bolts 1–10 in the order shown.

- Fit the cylinder head bolts and tighten them in the order shown in Fig. 1.41 in the following manner.
- Tighten the bolts in the order 1–10 to an initial torque of 2.2 kgm (16 ft.lb.).
- Tighten the bolts in the order 1–10 to a torque of 6.0 kgm (43.5 ft.lb.).
- Slacken all bolts in the order 1–10 by one complete turn.
- Tighten all bolts in the order 1–10 to a torque of 6.0 kg (43.5 ft.lb.).
- All bolts must now be angle-tightened by 220° (+/−5°). Either use a graduated disc or estimate the angle. One half of a turn represents 180°, and 1/8 of a turn represents 45°. Using these values where should be no problems to arrive at the correct angle.

- Fit the R.H. engine mounting and the cylinder head cover.
- Fit the camshaft positioning sensor by referring to Fig. 1.42. Fit the sensor (1) and the screw (2). The sensor is pushed to the end of its slot before the screw is tightened.
- Fit the camshaft gearwheel and fit the bolt (3) in Fig. 1.42. Tighten the bolt to 4.3 kgm (31 ft.lb.). The gearwheel must be prevented from rotating, as shown in Fig. 1.40.

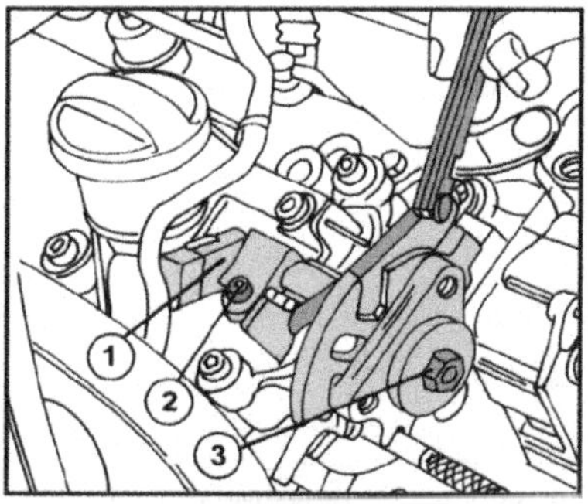

Fig. 1.42 – Adjusting the gap of the camshaft positioning sensor.

- The air gap for the positioning sensor must now be adjusted by referring to Fig. 1.42. The gap measured at the position shown must be set to 1.2 mm. Correct the gap by moving the sensor and then tighten the screw (2).
- Tighten the bolts (1) and (2) in Fig. 1.43.
- Refit the high-pressure injection rail as described in the relevant section. Also, fit the injectors.

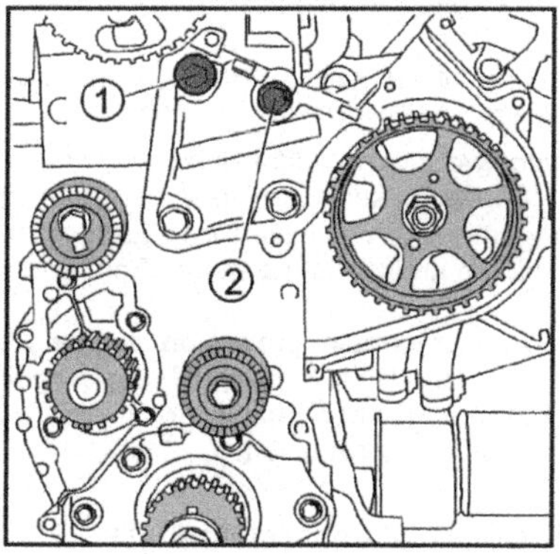

Fig. 1.43 – Tighten the bolts (1) and (2) at the positions shown.

- Fit the timing belt and the drive belt. Both operations are described later on.
- Fit the sound-proofing panel underneath the vehicle and refit the R.H. road wheel.
- Refit the coolant outlet housing with new seals, fit the studs and tighten the housing with the nuts. Re-connect the hoses to the outlet housing.
- Re-connect all connector plugs and refit the EGR valve with the heat exchanger.
- The remaining operations are carried out by referring to Fig. 1.27. Refit the support bracket (2) and the reservoir for the power steering fluid (1). Position and re-connect the vapour re-circulating hose (5), the fuel supply and return hoses (13) and the venting hoses (4). Fit the expansion tank (3), trim cover (6), the air inlet collar assembly (7) to the throttle valve housing, the air duct (12), the air filter (8), the air flow meter (10) and the air inlet duct (9). Fit the connector (11).
- Fill and bleed the cooling system as described in Section 1.4.5. (commencing page 36).
- Finally, re-connect the battery, lower the vehicle to the ground and tighten the wheel.

1.4.1. PISTONS AND CONNECTING RODS

1.4.1.0. Technical Data

Type:	Aluminium alloy pistons with 3 piston rings. With oil cooling jets for pistons. Floating fit piston pins in pistons and connecting rods. Pins held in position by snap rings.
Piston fitting direction	Side marked "Dist" must face the timing belt.
Piston Crown Protrusion Above Cylinder Block	See under "Cylinder Head"
Piston Diameters:	
- Standard:	84.931 mm
- Repair size:	85.531 mm
Piston running clearance:	0.062–0.078 mm
Piston Rings:	
- Type	2 compression rings, 1 oil control ring.
- Fitting direction	"Top" must be readable from above.
Piston Ring Thickness:	
- Upper rings:	3.5 mm
- Centre rings:	2.00 mm
- Oil control ring:	3.00 mm
Piston Ring gaps:	
- Upper and centre rings	0.20–0.35 mm
- Oil control rings	0.25–0.50 mm

- Gap arrangement	At 120° intervals
Piston Pins:	
- Outer diameter:	24.994–25.000 mm
- Length:	66.00 mm

Connection Rods and Big End Bearings

Type:	Forged steel, "I" section
Bearing shell material:	Aluminium-tin
Connecting rod length, centre to centre	152.0 mm
Big end bearing bore	53.695 mm
Small end bore:	25.00 mm
Big End Bearing Shell Thickness	
- Nominal	1.828 mm
- Repair size	1.978 mm

1.4.1.1. General

The pistons and connecting rods can only be replaced when the engine has been removed from the vehicle and the transmission disconnected from the engine.

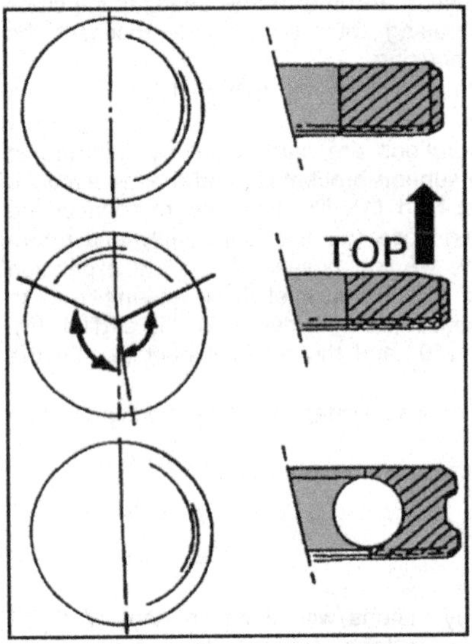

Fig. 1.44 – The arrangement of the piston rings. Note the fitting direction of the second ring and the arrangement of the ring gaps.

Each piston is fitted with two compression rings and one oil control ring. Fig. 1.44 shows a sectional view of a piston, together with the arrangement of the piston ring gaps. Note that the marking of the piston rings must be visible from above after installation.

The pistons must have a certain protrusion above the surface of the cylinder block. In the case of the engine covered in the manual, a cylinder head gasket of a different thickness is selected to obtain the protrusion.

1.4.1.2. Dismantling Pistons and Connecting Rods

No special tools are necessary to dismantle the pistons and connecting rods. Securing rings are fitted to hold the piston in position. Proceed as follows to dismantle. Fig. 1.46 shows an exploded view of the assembly.

- Use a pair of piston ring pliers and remove the piston rings one after another from the crown end of the piston, as shown in Fig. 1.45. If no piston ring pliers are available, use three thin metal strips and slide the rings over the strips to remove them one by one. One of the strips must be placed underneath the piston ring ends to avoid scratches.

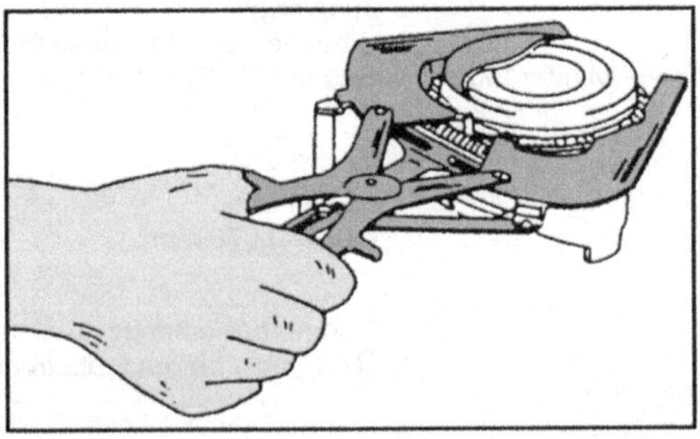

Fig. 1.45 – Removal or installation of piston rings.

- Remove the securing rings out of the two piston pin bores. Either use a small screwdriver or a pair of circlip pliers for this operation. The circlips may be tight.

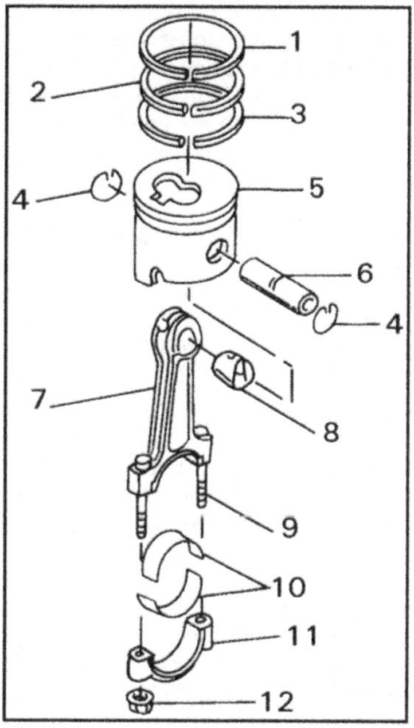

Fig. 1.46 – Exploded view of a piston and connecting rod assembly.

1 Upper piston ring
2 Centre piston ring
3 Oil control ring
4 Piston pin retaining ring
5 Piston
6 Piston pin
7 Connecting rod
8 Big end bearing bush
9 Big end bearing bolt
10 Big end bearing shells
11 Big end bearing cap
12 Big end bearing cap nut
13 Crankshaft

- Use a stepped mandrel to drive out the piston pin. Hold the piston in one palm of the hand, whilst the pin is removed. If the piston has a tight fit, heat the piston in water.

Note: Pistons and connecting rods should only be separated if pistons or connecting rods must be replaced. Experience has shown that it is of advantage to have a complete piston set replaced in an engine shop. There should, however, be no problems to replace a single piston.

1.4.1.3. Checking Pistons and Connecting Rods

Thoroughly clean and check all parts. Signs of scratches or seizure or wear mean that the respective parts must be replaced.

- One after the other, insert each piston ring into the top of the cylinder bore (from where it was removed) and push it squarely into the bore, using an upside-down piston to do so, until the ring is inserted by approx. 15–20 mm. Insert a feeler gauge into the gap between the two ring ends and check that the gap obtained is not greater than specified in Section 1.4.1.0. Ring gaps cannot be adjusted and new rings must be fitted, if the gaps of any of the rings are greater. Fig. 1.47 demonstrates the check.
- Measure the piston diameter with a micrometre. Apply the jaws of the micrometre at right angles to the piston pin bore at the bottom of the piston skirt. One oversize piston size is available and we advise you to seek assistance from an engine shop, dealing with Citroen/Peugeot engines.

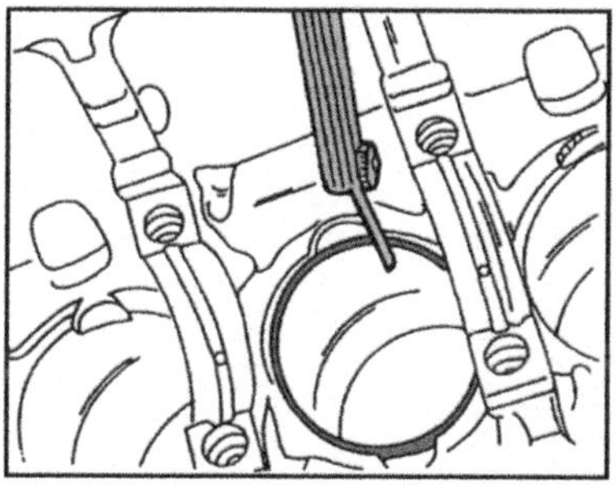

Fig. 1.47 – Checking piston ring gaps.

- Check the piston pins and the bores in the pistons for wear. Piston pins must have a sliding fit in the pistons and connecting rod small ends, with the parts well-oiled.

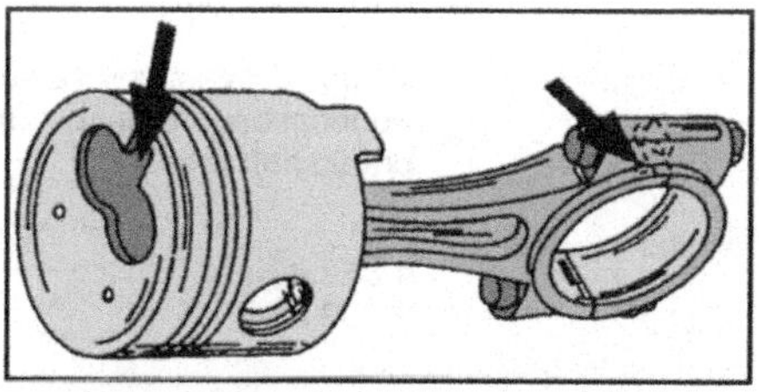

Fig. 1.48 – Correct installation of piston and connecting rod.

1.4.1.4. Assembling Pistons and Connecting Rods

It is assumed that the parts have been checked as described in the last section and parts have been replaced, if necessary.

Lubricate the piston pin bores and the connecting rod small end bores and assemble the parts correctly, i.e., the cut-out in the piston bowl must be on the same side as the locating tabs of the connecting rod bearing shells, as shown in Fig. 1.48.

Fit one of the piston pin securing cringes into the piston pin bore. Check that the ring is engaged fully in its groove.

Insert the connecting rod into the piston, align the small end bore and the piston bore and insert the piston pin, using thumb pressure only. Check once more the correct relationship of piston and connecting rod and fit the second securing ring to the other side of the piston. Check the securing ring for proper fit.

Fit the piston rings over the pistons, using a pair of piston ring pliers, as shown in Fig. 1.45. The centre ring must be fitted so that the marking "Top" can be read from above after the rings have been fitted. If a piston ring compressor is not available, use three thin metal strips. Slide the rings to the level of the ring groove and then withdraw the strips. The ring will spring into the groove. Piston rings break easily and care should be taken.

Generously lubricate the piston rings and arrange the piston ring gas in equal spacing around the outside of the piston skirt. Refer to Fig. 1.44. Arrange the gaps as shown.

Lubricate the pistons with engine oil and place the assemblies for installation in the cylinder block. A piston ring compressor is required to fit the pistons.

The installation of the pistons and connecting rods is described in detail in Section 1.3. Tightening torques are given at the end of this chapter.

1.4.2. CYLINDER BLOCK

The cylinder block consists of the crankcase and the actual block with the cylinder bores. Special attention should be given to the cylinder block at each major overhaul of the engine, irrespective of whether the bores have been re-machined or not. Thoroughly clean all cavities and passages and remove all traces of foreign matter from the joint faces. If any machining of the bores has taken place, it is essential that all swarf is removed before assembly of the engine takes place.

Measurement of the cylinder bores should be left to an engine shop, as they have the proper equipment to measure cylinder bores. It is, however, feasible to check the cylinder block face for distortion, in a similar manner as described for the cylinder head. The max. permissible distortion is 0.05 mm. As already mentioned, there is no provision to have the block surface re-machined.

1.4.3. CRANKSHAFT AND FLYWHEEL

1.4.3.0. Technical Data

Number of main journals	5
Main Bearings:	
- Type	Steel type
- Material	Aluminium-tin
Main Bearing Diameter:	
- Nominal diameter	60.000 mm
- Repair size diameter	57.700 mm
Crankpin Diameter:	
- Nominal diameter	50.000 mm
- Repair size diameter	49.700 mm
Crankshaft end float	0.07–0.32 mm
Thickness of thrust washers:	2.305, 3 oversizes (2.40, 2.45 and 2.50 mm). Lubrication groove must face towards crankshaft face.
Thickness of Main Bearing Shells:	
- Nominal thickness	1.853 mm
- Repair size thickness	2.003 mm

1.4.3.1. General

The removal of the crankshaft is described in Section 1.2. The crankshaft runs in five bearings. Main bearing and crankpin journals can be re-grounded once, i.e., repair size bearing shells are available. Fig. 1.49 shows the component parts of a crankshaft.

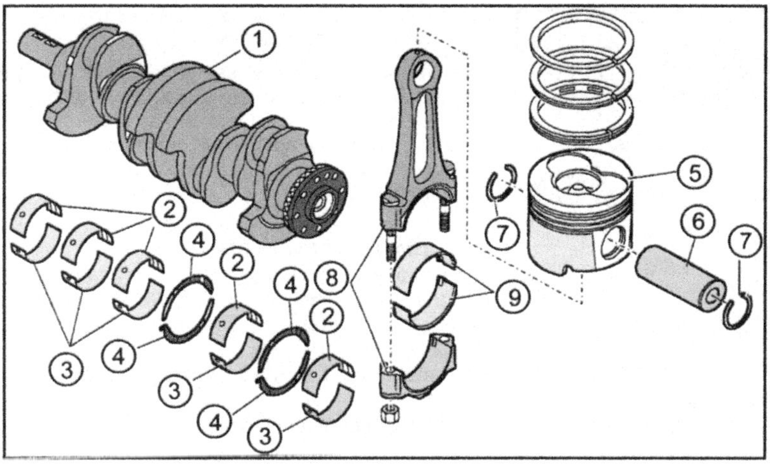

Fig. 1.49 – Crankshaft and piston assembly.

1	Crankshaft	4	Thrust half washers	7	Circlips
2	Upper bearing shells	5	Piston	8	Connecting rod
3	Lower bearing shells	6	Piston pin	9	Big end shells

An oil seal is fitted to the front and rear ends of the crankshaft. The front seal is inserted into a removable cover. The rear oil seal is fitted to the crankcase and the rear main bearing cap.

1.4.3.2. Checking the Crankshaft End Float

Check the end float of the crankshaft before it is removed from the crankcase. The resulting valve can then be used to correct the end float during installation of the shaft, by fitting oversize thrust washers. Check the end float as follows.

- Attach a dial gauge with a suitable bracket to the cylinder block end face, as already shown during the assembly of the engine and place the dial gauge stylus onto the shaft end. Push the crankshaft to one side, using a screwdriver and set the dial gauge to "Zero".
- Push the crankshaft into the other direction and read off the value on the dial gauge. Check that the end float is within the values given in Section 1.4.3.0. If the end float is outside the upper limit, fit oversize thrust washers. Upper and lower thrust washers, however, must always have the same size.

1.4.3.3. Crankshaft – Installation

The installation of the crankshaft has been described during the assembly of the engine. As already mentioned, always use two thrust washers of the same thickness.

1.4.3.4. Rear Crankshaft Oil Seal – Replacement

The flywheel must be removed to replace the oil seal. A slipping clutch, for example, is a sign of a leaking oil seal.

To replace the oil seal without damaging any parts, screw a self-tapping screw into the outside of the seal and use a screwdriver as lever to lever out the seal. Check the running area of the crankshaft for wear. If wear is visible, drive the new oil seal in a little further.

Fig. 1.50 – Fitting the rear crankshaft oil seal. Place a plate (1) over the seal and drive it into the cylinder block.

Coat the outside of the new seal with engine oil and the seal lip with grease and drive the seal into the crankcase as shown in Fig. 1.50. A flat plate as shown in the illustration is ideal to drive the seal in position. Drive the seal in position until flush, or a little further if the crankshaft has shown signs of wear. Check that the oil seal is fitted evenly on its entire circumference.

1.4.3.5. Front Crankshaft Oil Seal – Replacement

The front crankshaft oil seal is fitted to the flange in front of the oil pump and can be replaced with the engine fitted. The timing belt must be removed. Remove the old oil seal by screwing two self-tapping screws into the seal end face and pull it out with the help of a screwdriver and a pair of pliers.

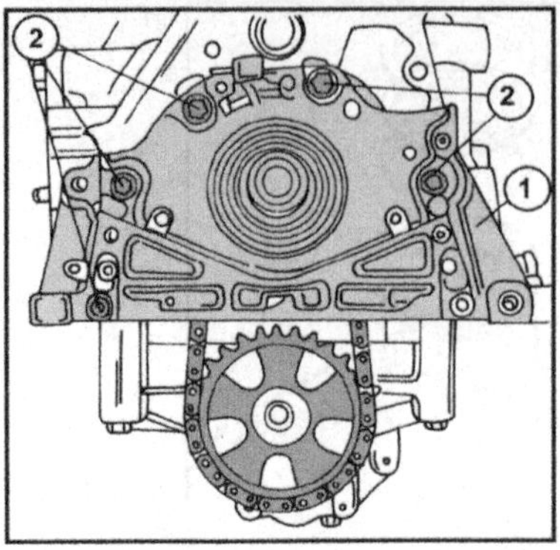

Fig. 1.51 – The front flange (1) is secured by screws (2). Tighten the screws to 1.2 kgm after installation.

Coat the lip of a new oil seal with a little MP grease and the outside with engine oil. Drive the seal into the cover until the outside is flush. A plate, the same as shown in Fig. 1.50, can be used to install the seal. Refit the front cover by referring to Fig. 1.51. Finally, refit the removed parts. Installation of the timing belt follows below.

1.4.4. TIMING MECHANISM

The component parts of the timing mechanism can be removed with the engine fitted to the vehicle. The belt must be partially removed, i.e., from the camshaft gear, when the following operations are necessary.

- Removal and installation of the cylinder head.
- Removal and installation of the camshaft.
- Removal and installation of the oil pump drive chain or drive sprocket.
- Removal and installation of the water pump.

1.4.4.1. Removal and Installation of the Timing Belt

A special instrument to check the timing belt tension is required to check the tension after installation of the belt. If you have, however, some experience with engines, you will be able to measure the tension of the belt before removal. A spring scale is required. Hook the spring scale on the timing belt and pull the scale towards the outside. Write down the force required to pull out the belt. During later installation, tension the belt until the same spring scale indication is obtained.

As the engine mounting must be removed, you will need a hand crane or other lifting tackle to lift up the engine. You also need a locating peg for the flywheel and an arrester pin for the camshaft gearwheel. The tensioning of the timing belt is also carried out by means of a special tool. This will be shown later on and necessary make-shift tools are described. Replace the timing belt as follows.

- Jack up the R.H. front of the vehicle, remove the front wheel and place a chassis stand underneath the body. Engage the highest gear to be able to rotate the engine during later operations. Disconnect the battery.
- Remove the protector inside the front wheel arch and the sound-insulating panel underneath the engine.

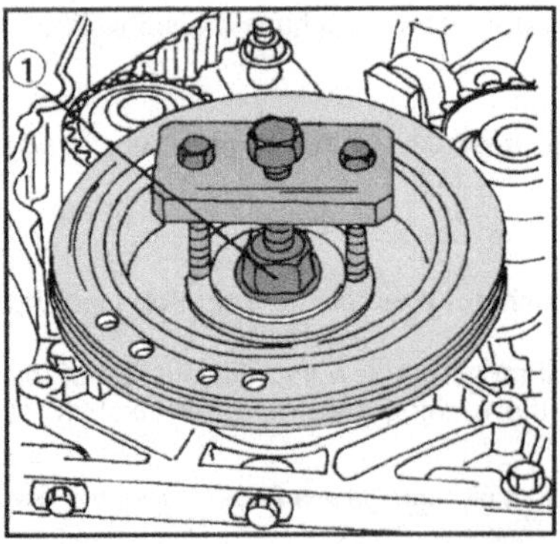

Fig. 1.52 – Removal of the crankshaft pulley. The puller is fixed to the pulley as shown. By tightening the centre bolt, the pulley will come away.

Slacken the drive belt at the front of the engine and remove it. See the description in Section 1.4.5. (commencing page 36).

- Remove the closing cover underneath the flywheel housing. Lock the flywheel in a suitable manner. Fig. 1.14 shows a suitable method. In our case, the "flywheel lock" will have to be inserted at the bottom.
- Remove the crankshaft pulley bolt (flywheel locked as above) and withdraw the pulley. The pulley has a tight fit and a puller, as shown in Fig. 1.52, can be used to remove it.
- Attach a rope or chain to the engine lifting brackets and lift the engine until the engine mounting is under tension. It is also possible to place a mobile jack underneath the engine (wooden plank for protection).

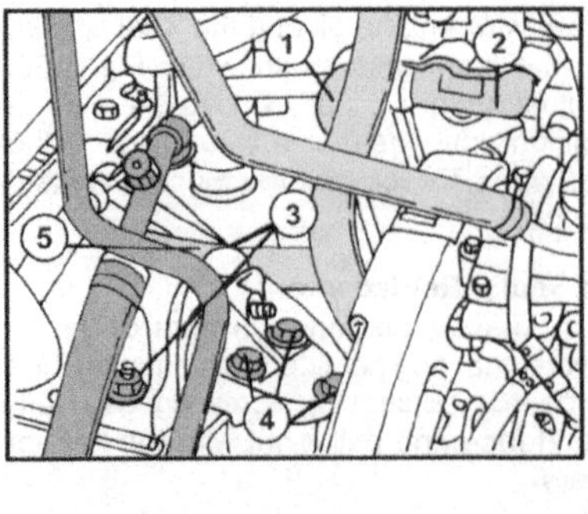

Fig. 1.53 – Details for the removal and installation of the timing belt. The numbers are referred to the text.

- Remove the engine mounting on the side of the timing belt by referring to Fig. 1.53. Remove in the following order the heat shield (1), the pipe (2), the nuts (3), the bolts (4) and the R.H. engine mounting (5).
- Lift and lower the engine to gain access to all bolts securing the timing belt covers.

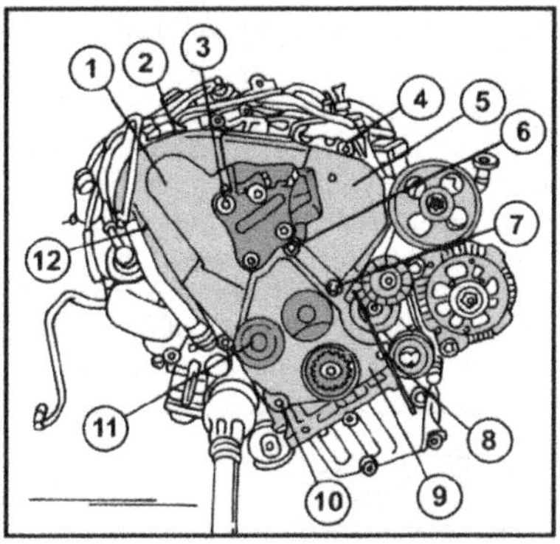

Fig. 1.54 – Details for the removal and installation of the timing belt covers. The numbers are referred to the text.

- Refer to Fig. 1.54 and remove the parts shown in the illustration. The engine is removed for clarity. Remove the bolts (2), (3), (11) and (12) and take off the upper timing belt cover (1). Also, remove the side cover (5) after removal of the bolts (6) and (7). Remove bolts (8) and (10) and take off the lower timing belt cover (9). Note that the bolt (11) also secures the water pump. After removal of the bolt, find a spacer ring with a thickness of 17 mm and refit the bolt with the spacer ring. Tighten the bolt to 1.5 kgm (11 ft.lb.).

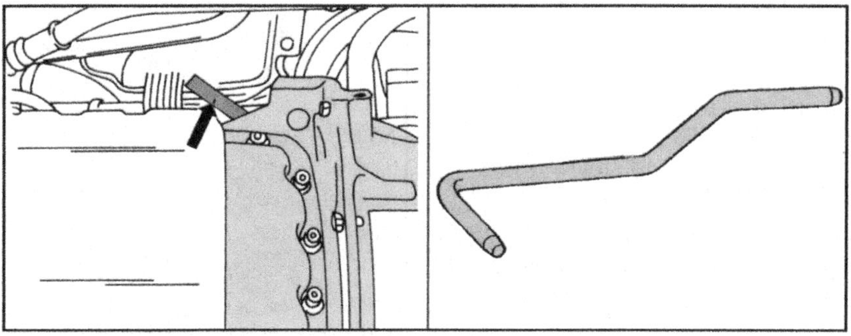

Fig. 1.55 – To lock the flywheel in position, insert the timing rod shown on the R.H. side into the cylinder block, at the position shown.

- With the gear engaged, turn the wheel hub of the jacked-up front suspension until the piston of No. 1 cylinder is at the top dead centre position, i.e., the timing rod must be inserted into the flywheel/crankshaft at the position shown on the L.H. side of Fig. 1.55. Rotate the crankshaft to and fro until you can feel the engagement of the timing rod. The shape of the timing rod is shown in the R.H. view.

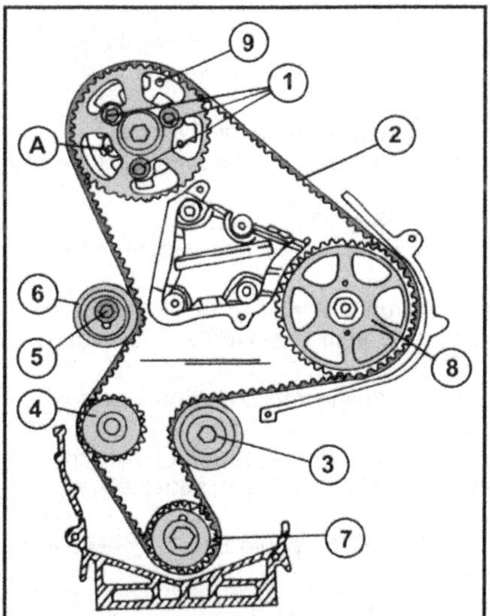

Fig. 1.56 – Details for the removal and installation of the timing belt. The numbers are referred to the text.

- The next operations are carried out by referring to Fig. 1.56. First, insert a bolt of suitable diameter into the bore of the camshaft timing gear to lock the gear in position. Slacken the three bolts (1) securing the timing gear and the bolt (5) securing the tensioning roller (6). Push the tensioning roller towards the outside and remove the now slack timing belt (2). If the old belt is to be refitted, mark the running direction into the outside of the belt. Use a felt pen and mark an arrow into the belt. If a new belt is fitted, it is advisable to replace the roller (6) and the guide roller (3). Remove the bolt in the centre to remove the guide roller.
- All parts of the timing mechanism can now be replaced as necessary.

The installation of the timing belt is carried out as follows. The numbers are referred to Fig. 1.56.

- Fit the bolts (3) into the camshaft timing gear and rotate the gearwheel towards the right until the bolts are in contact with the ends of the elongated holes.
- Place the timing belt (2) over the crankshaft timing gear (7). Make sure that the belt cannot slip off the gearwheel.

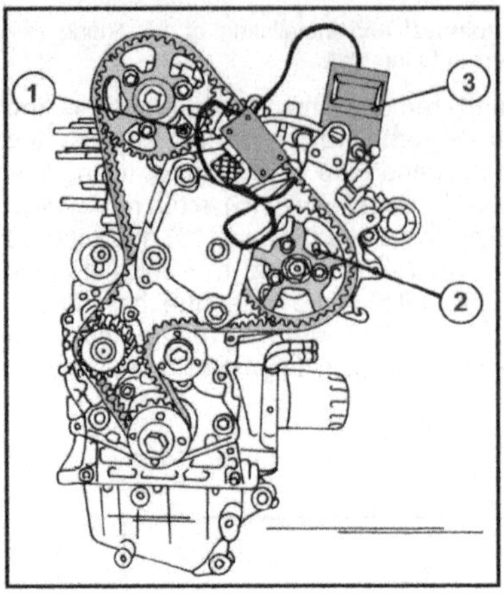

Fig. 1.57 – Checking the belt tension with the special checking tool.
1 Camshaft gear lock pin
2 HP pump drive gear
3 Tension measuring gauge

- Fit the timing belt in the following order: guide roller (3), drive wheel for the HP pump (8), camshaft timing gear (9), water pump drive gearwheel (4) and the belt tensioning roller (6). If necessary, rotate the camshaft timing gear slightly towards the left to make sure the belt engages with all teeth of the various sprockets. Check once more that all teeth have engaged.

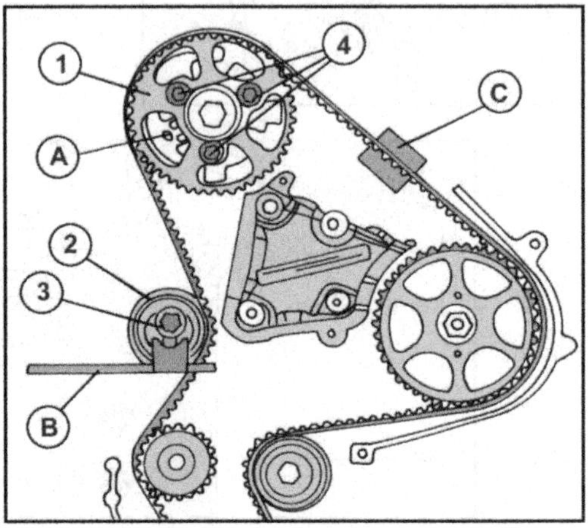

Fig. 1.58 – Details for the installation of the timing belt. The letters and numbers are referred to the text.

In the workshop, a special tool is used to tension the timing belt in the workshop. This is placed against the long side of the timing belt, as shown in Fig. 1.57, and shows the tension of the belt in SEEM units. If the tool cannot be obtained, you can adjust the belt tension with the instructions given during the removal of the timing belt, i.e., the tension of the belt is measured with a spring scale. With some experience with timing belts, there should be no problems. Tension the belt by referring to Fig. 1.58.

- Fit the tensioning measuring instrument shown in Fig. 1.57 (3) to the timing belt as shown in Fig. 1.58 (indicated by "C") and slacken the bolt (3) securing the tensioning roller (4). Using the adjusting tool (B), turn the tensioning roller (2) towards the left (anti-clockwise) until a reading of 98 SEEM +/− 2 SEEM is indicated. Item (3) in Fig. 1.57 shows where the instrument is positioned. In this position, tighten the bolt (3) in Fig. 1.58 to 2.5 kgm (18 ft.lb.).
- Check that the bolts securing the camshaft gearwheel are not contacting the ends of the elongated holes. To do this, remove one of the bolts (4) from the camshaft gearwheel (1) and check the position of the threaded bore. If the position is in order, refit the bolt and then re-tighten all bolts to 2.0 kgm (14.5 ft.lb.). If this is not the case, remove the belt and start from the beginning.
- Remove all tools indicated by letters in Fig. 1.58 and withdraw the lock rod out of the flywheel.
- Rotate the crankshaft by exactly 8 revolutions in the normal direction of rotation and re-insert the lock rod into the flywheel, as shown in Fig. 1.55.
- Slacken the bolts (4) and insert the locking pin in its place into the camshaft gearwheel.
- Slacken the bolt (3) to free the tensioning roller (4). Refit the tension measuring instrument to the timing belt.
- Turn the tensioning roller (2) with the adjuster (B) anti-clockwise until a reading of 54 SEEM +/− 2 SEEM is obtained. In this roller position, tighten the bolt (3) to 2.5 kgm (18 ft.lb.) and the bolts (4) securing the camshaft timing gear to 2.0 kgm (14.5 ft.lb.).
- Remove the locking pin out of the camshaft gearwheel and the timing rod out of the rear of the crankcase. Also, remove the measuring instrument from the belt.
- Turn the crankshaft by exactly 8 revolutions in the direction of rotation and re-insert the timing rod into the flywheel. If it can be inserted without difficulties, also insert the timing pin into the camshaft gearwheel.
- Refit the measuring instrument to the belt and check the tension, which should now be between 40 and 54 SEEM +/− 3 SEEM. If this is the case, remove all tools.

The remaining operations are now carried out as follows. Details are shown in Fig. 1.54.

- Unscrew the bolt (11) and the spacer.
- Fit the lower timing belt cover (9) and secure it with bolts (8) and (10).
- Fit the centre timing belt cover (5) and secure with bolts (3), (6) and (7).
- Fit the upper timing belt cover (1), fit the bolt (11)–1.5 kgm (10 ft.lb.) and insert and tighten the bolts (2), (3) and (12) to secure the belt cover.

The parts in the engine compartment are now fitted as follows by referring to the details in Fig. 1.53.

- Fit the R.H. engine mounting (5) and fit and tighten the bolt (4) and the nut (3).
- Fit the pipe (2) and the heat shield (1). The lifting tackle can now be removed.
- Coat the crankshaft pulley bolt with "Loctite" or a similar agent, fit the crankshaft pulley over the crankshaft and fit the bolt with a washer. Tighten the bolt to 4.0 kgm (29 ft.lb.). The bolt is now angle-tightened by a further 50° without using the torque wrench. A tolerance of 5° plus or minus is permissible, i.e., 1/8 of a turn is acceptable.
- All remaining operations are carried out in reverse order.

1.4.4.2. Removal and Installation of Camshaft

The cylinder head can remain fitted to the engine if only the camshaft is removed. Proceed as follows.

- Disconnect the battery earth cable. Refer to Fig. 1.59 and remove the air duct (1), the air cleaner (2), the air flow meter (3) and the connector (4). Referring to the R.H. view of the illustration removes the EGR valve (5) (exhaust gas return valve), the heat exchanger (6) and the venting tube (7).
- Remove the drive belt from the front end of the engine as described later on. The R.H. engine mounting, removed for the operation, must be subsequently refitted.

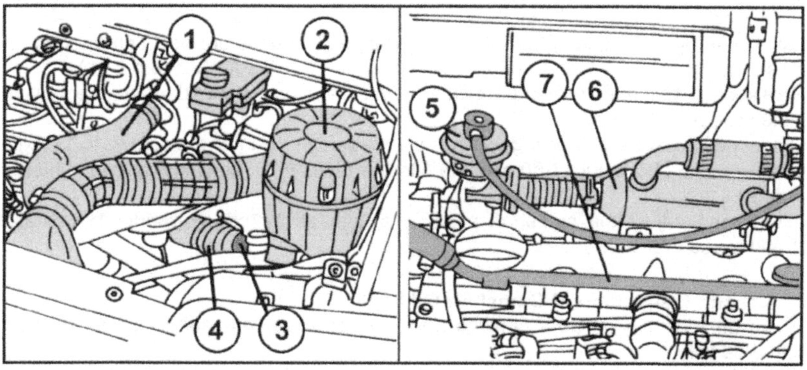

Fig. 1.59 – Details for the removal and installation of the crankshaft. The items (1)–(7) are removed in the engine compartment (see text).

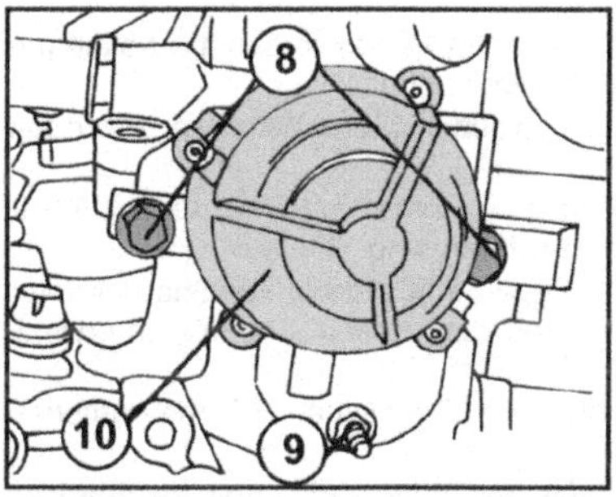

Fig. 1.60 – The vacuum pump (10) is attached at the position shown. A screw (8) and a nut (9) are used to secure the pump.

- Remove the bolts (8) in Fig. 1.60 and the nut (9) and take off the vacuum pump (10).
- Prevent the camshaft gearwheel from rotating as already shown in Fig. 1.40 and slacken the bolt in the centre of the gearwheel. Withdraw the gearwheel from the end of the camshaft. Remove the cylinder head cover (Fig. 1.30).

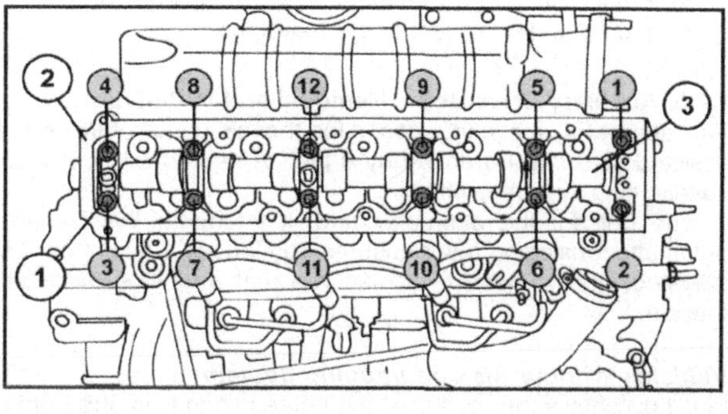

Fig. 1.61 – Slacken the camshaft housing securing bolts (1) in the order 1–12 to remove the housing (2). The camshaft (3) remains in the housing.

- The camshaft can now be removed by referring to Fig. 1.61.
- Slacken the bolts (1) in Fig. 1.61 in the order 1–12 to remove the camshaft housing (2) until all bolts are free to be removed by hand. Grip the camshaft housing as shown in Fig. 1.62 and lift it off the cylinder head. The camshaft can now be removed from the housing. Remove the gasket between camshaft housing and cylinder head. Thoroughly clean the gasket faces with a suitable cleaning fluid. There must be do damage to any of the gasket faces. Do not use emery cloth or similar to clean them.

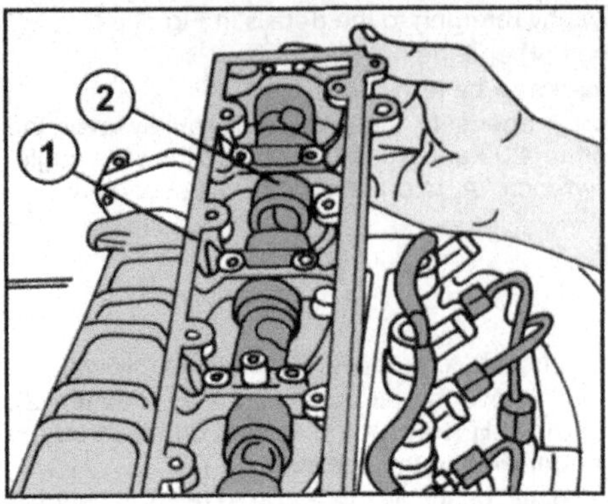

Fig. 1.62 – Removal of the camshaft housing (1) together with the fitted camshaft (1).

The installation is carried out as follows.

- Once more check the gasket faces for signs of foreign matter.
- Rotate the crankshaft until all pistons are approximately in the centre of their stroke. The Woodruff key in the crankshaft end should be in a horizontal position.
- Coat the sealing face of the camshaft bearing housing with sealing compound. You will have to find out in a workshop which type of sealant is used. In the case of Citroen, this is sealant available under the type designation "E6".
- Coat the camshaft bearing caps and cams with engine oil and insert the shaft into the housing. Place the camshaft housing over the cylinder head and loosely fit the bolts.
- Fit the bolts (1) in Fig. 1.61 and tighten them finger-tight. The bolts are now tightened in reverse order to the one shown in the illustration to a torque reading of 1.0–1.2 kgm (7.5–8.5 ft.lb.).
- Fit a new oil seal as described under the next heading. Take care not to damage the sealing lip. Do not lubricate the outside of the seal.
- Fit the cylinder head cover with a new gasket and fit the screws. Tighten the screws to 1.0 kgm (7.5 ft.lb.).
- Fit the vacuum pump (10) in Fig. 1.60 with the filter and a new "O" sealing ring and tighten the bolts (8) and the nut (9).
- Finally, check and adjust the camshaft position sensor as described during the installation of the cylinder head (page 27).

1.4.4.3. Removal and Installation of Camshaft Oil Seal

The oil seal can be replaced with the camshaft housing and the camshaft in position, i.e., fitted to the engine. Proceed as follows.

- Disconnect the battery earth cable.
- Remove the drive belt from the front of the engine as described later on.
- Remove the timing belt as already described.
- Prevent the camshaft gearwheel from rotating as already shown in Fig. 1.40 and remove the bolt in the centre of the gearwheel. Take care not to damage the camshaft position sensor.
- Raise the engine either with a lifting tackle or carefully with a jack (wooden plank between jack head and engine) to improve the access to the oil seal.
- Drill a hole into the outside of the seal and screw in a self-tapping screw into the seal. With a screwdriver and a pair of pliers, remove the oil seal out of the camshaft housing.
- Thoroughly clean the inside of the seal bore and fit the new seal. A piece of tube of suitable diameter can be placed against the seal to drive it in position. The outside of the seal must not be lubricated with engine oil.
- Adjust the air gap of the camshaft position sensor as described on page 27 and then refit the camshaft gearwheel. The bolt is tightened to 4.3 kgm (31 ft.lb.).
- All other operations are carried out in reverse order. Timing belt and drive belt must be fitted as described under the respective headings.

1.4.5. Auxiliary Drive Belt – Removal and Installation

The alternator, the compressor for the air conditioning system (if fitted) and the pump for the power-assisted steering are driven by a belt. A separate belt is used for the power steering pump when an air conditioning system is fitted.

The belt for the alternator and, if fitted, for the compressor, does not require adjustment, as an automatic tensioning roller ensures continuous adjustment. The belt for the power steering pump must be adjusted, but we have to advise you that a tension measuring instrument is used to obtain the correct tension.

Vehicles Without Air Conditioning System

Fig. 1.63 shows the routing of the belt as seen from the front of the engine. The belt tensioner (3) and the two jockey pulleys are rotating on bearing spigots. Before fitting a new belt, make a point to check all three rollers for free, unrestricted rotation. They must rotate without hard spots and must not have excessive play. If in doubt, fit a new roller or replace them all.

To reset the tensioning roller (1), use a ring spanner of suitable size which is applied to the bolt "b" of the roller. To retain the roller in its position after it has been released, you will have to lock it in its new position. This requires a drill bit or a metal pin (for example, the end of a suitable drift) which is inserted into the bore "a" in Fig. 1.63 next to the tensioning roller.

Replace the drive belt as follows.

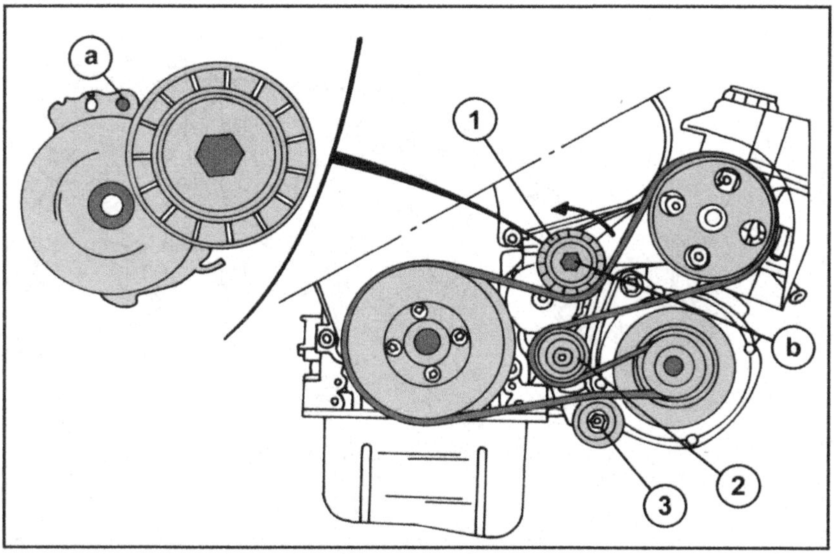

Fig. 1.63 – Details for the removal and installation of the auxiliary drive belt if the vehicle has no air conditioning system. The letters and numbers are explained in the text.

- Place the front end of the vehicle on chassis stands, remove the sound insulation panel underneath the engine and remove the R.H. front wheel.
- If the belt is to be refitted after removal, mark the outside of the belt with a felt pen to indicate the running direction of the belt. Make sure that the belt is refitted correspondingly.
- Place a ring spanner over the bolt "b" and rotate the bolt towards the left until the 4-mm pin/drift mentioned above can be inserted into the hole "a", as soon as the hole appears behind the pulley/roller. The tensioning roller will now remain in this position and the slack belt can be removed.
- Fit the belt in accordance with the diagram in Fig. 1.63. Check that the belt has engaged in the grooves of all pulleys and rollers.
- Apply the ring spanner to the bolt "b" in the centre of the roller (1) and rotate the roller until the 4-mm drill/drift can be withdrawn. Slowly release the ring spanner until the roller is pressing against the belt. The belt is now properly tensioned.

Vehicles With Air Conditioning System

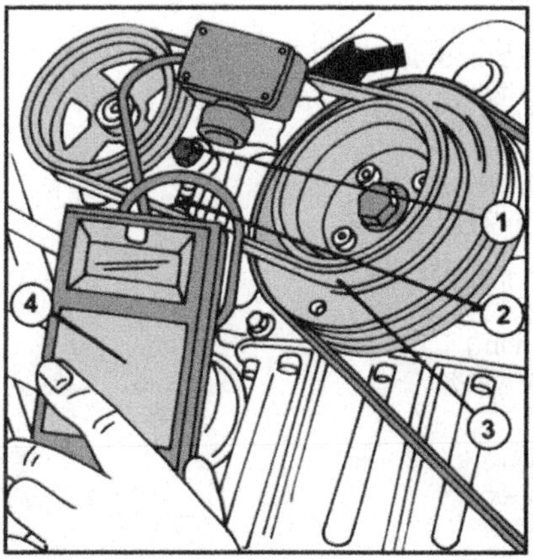

First, remove the belt for the power steering by referring to Fig. 1.64.

- Prepare the vehicle as described above. If the belt is to be refitted, mark the running direction into the outside, using a felt tip pen.
- Slacken the bolt (1), unscrew the nut (2) as far as necessary and remove the belt (3) as soon as it is slack and free.

During installation, place the belt into the grooves of the two pulleys and adjust the tension as follows.

Fig. 1.64 – The auxiliary drive belt in the case of a vehicle with air conditioning system. The measuring instrument is placed against the belt at the position shown by the arrow.

1 Lock bolt
2 Nut to adjust belt tension
3 Drive belt
4 Measuring instrument

- Place the measuring instrument against the belt on the upper run and tighten the nut (2) until a reading of 102 SEEM +/− 10 SEEM has been obtained. Keep this tension and tighten the bolt (1).
- Rotate the crankshaft by three revolutions in the normal direction of rotation and re-check the tension. The same value should be obtained.
- **If the measuring instrument is not available,** adjust the tension until the belt can be deflected by 4–5 mm.

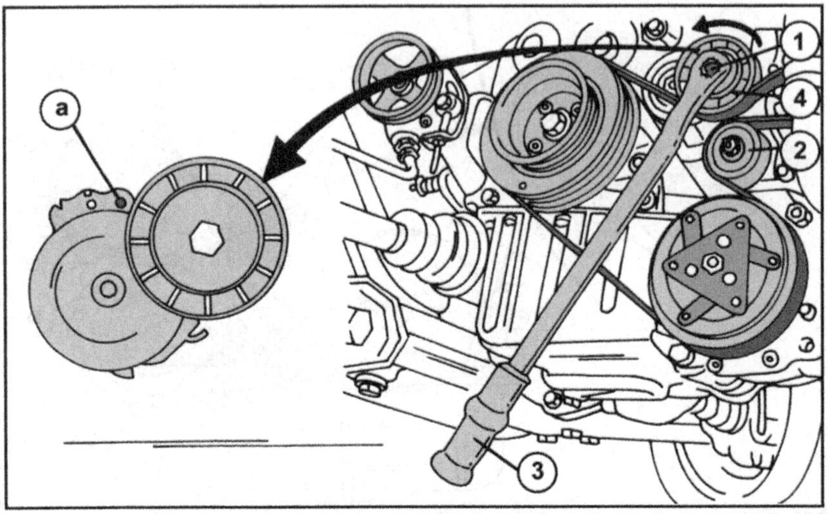

Fig. 1.65 – Details for the removal and installation of the drive belt for the alternator and the compressor if an A/C system is fitted.
1 Bolt for belt tensioner   2 Belt guide roller   3 Spanner   4 Belt tensioning roller

Remove the drive belt for the compressor by referring to Fig. 1.65.

- Turn the bolt (1) with a socket and an extension in an anti-clockwise direction until the bore "a" is visible as shown. Insert a drift or drill bit of 4 mm in diameter into the bore and allow the tension roller (4) to return.
- Remove the belt from the pulleys. Mark the running direction of the belt into the outside, using a felt tip pen, if the same belt is installed.

The installation is carried out in reverse order. Before fitting the belt, check that the tensioning roller (4) and the guide roller (2) can be rotated without hard spots or excessive play. If in doubt, replace the roller in question or both. The drift/drill bit can be removed after the tensioning roller has been released.

1.5. Tightening Torque Values – Engine

Cylinder Head Bolts:	
- 1st stage, initial tightening	2.2 kgm (16 ft.lb.)
- 2nd stage	6.0 kgm (43.5 ft.lb.)
- 3rd stage	Slacken all bolts in order by 1 turn
- 4th stage	6.0 kgm (43.5 ft.lb.)
- 5th stage	Tighten by 220° (+/−5°)
Cylinder head cover	10.0 kgm (7.2 ft.lb.)
Camshaft bearing caps	1.0 kgm (7.2 ft.lb.)
Camshaft timing gear centre bolt	4.3 kgm (31 ft.lb.)
Camshaft timing gear to hub (3 off)	2.0 kgm (14.5 ft.lb.)
Hub for camshaft timing gearwheel	2.0 kgm (14.5 ft.lb.)
Timing belt covers	1.0 kgm (7.2 ft.lb.)
Exhaust manifold	2.0 kgm (14.5 ft.lb.)
Main Bearing caps:	
- 1st stage	2.5 kgm (18 ft.lb.)
- 2nd stage	60° angle-tightening
Connecting Rod Bearing Caps:	
- 1st stage	2.0 kgm

- 2nd stage	70° angle-tightening
Flywheel bolts	5.0 kgm (36 ft.lb.)
Clutch bolts	2.0 kgm (14.5 ft.lb.)
Crankshaft pulley:	
- 1st stage	4.0 kgm (30 ft.lb.)
- 2nd stage	51° angle-tightening
Oil splash jet to cylinder block	1.0 kgm (7.2 ft.lb.)
Timing belt tensioning/guide roller	2.5 kgm (18 ft.lb.)
Oil pressure switch	3.0 kgm (22 ft.lb.)
Oil drain plug (sump)	3.0 kgm (22 ft.lb.)
Water drain plug (block)	2.5 kgm (18 ft.lb.)
Water pump	1.5 kgm (11 ft.lb.)
Water inlet elbow	2.0 kgm (14.5 ft.lb.)
Oil pump bolts	1.3 kgm (9.5 ft.lb.)
Oil cooler	5.8 kgm (42 ft.lb.)
Lubrication pipes for turbo charger:	
- Engine side	3.0 kgm (22 ft.lb.)
- Turbo charger side	2.0 kgm (14.5 ft.lb.)
R.H. engine mounting	2.7 kgm (20 ft.lb.)
Diesel Injection	
- Injector mounting flange (nut)	3.0 kgm (22 ft.lb.)
- Union nuts on HP injector rail	2.0 kgm (14.5 ft.lb.)
HP pump to mounting	2.25 kgm (16 ft.lb.)
Connector on diesel injector	2.2 kgm (16 ft.lb.)
HP pump pulley	5.0 kgm (36 ft.lb.)
Connector in HP pump	2.2 kgm (16 ft.lb.)

1.6. Lubrication System

1.6.0. TECHNICAL DATA

Oil sump capacity	Refer to page 6
Oil type (recommended)	Total oil for diesel or turbo diesel engines (10W40 or 15W40)
Oil Pressure:	
- Min. oil pressure at 1000 rpm	2.0 bar (28.5 psi.)
- Oil pressure at 80 °C and 4000 rpm	4.0 bar (57 psi.)
Oil Filter:	
- Type	Replaceable filter cartridge (Purflux)
- Changing intervals	Together with oil change or once a year (every 10 000 miles)

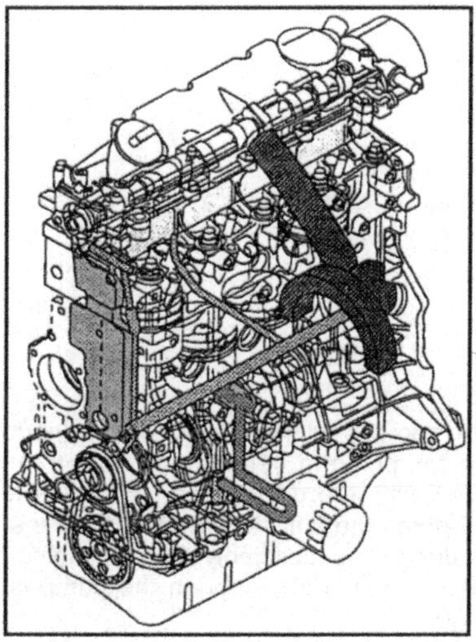

Fig. 1.66 – Sectional view of the engine to show the lubrication system.

1.6.1. GENERAL

An oil pump, driven by a separate chain from the crankshaft, supplies the oil for the lubrication of all bearing points and moveable parts. Note the following.

- The bottom of the pistons is cooled by oil splash jets. The jets are bolted to the cylinder block. The hydraulic valve clearance compensating elements are also supplied with oil from the lubrication system.
- The engine has an oil cooler between the oil filter and the cylinder block.
- Replace the oil filter at every oil change.

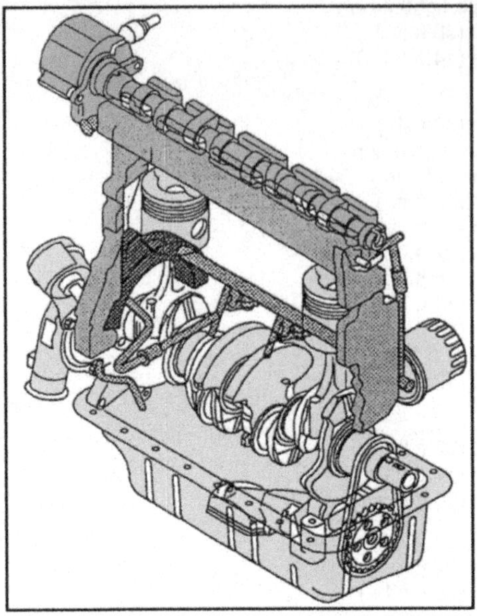

Fig. 1.67 – Sectional view of the engine to show the lubrication circuit.

- Note the oil capacity in the oil sump (page 6).
- The oil pressure is different at 1000 and 4000 rpm (see Section 1.6.0.).
- The oil sump is either made of steel (without A/C system) or aluminium (with A/C system).

Fig. 1.66 shows a sectional view of the lubrication circuit. Fig. 1.67 shows a view from the other side. Both illustrations show the location of the oil pump.

1.6.2. OIL PUMP (AND OIL SUMP)

The oil pump can be replaced with the engine fitted.

- Lift up the front end of the vehicle (chassis stands). It would, however, be easier if the vehicle can be driven onto a lifting ramp.
- Remove the protective panel underneath the engine.
- Drain the engine oil into a suitable container. Slightly warm oil will drain easier.
- Unscrew the oil sump securing screws and lower the sump. A sticking oil pump can be carefully freed by tapping it with a plastic or rubber mallet. Do not insert a screwdriver into the joining faces. Note the differences between a steel sump and an aluminium sump (see dismantling procedures for the engine).
- To remove the oil pump, refer to Section 1.2., covering the dismantling of the engine. The oil pump must be slightly turned to disengage it from the drive chain.

Repair of the oil pump is not possible. Replace the pump if a low oil pressure is the reason for the pump removal. Only the pump suction filter can be replaced. Fig. 1.68 shows the component parts of the oil pump and the oil sump and other parts of the lubrication system.

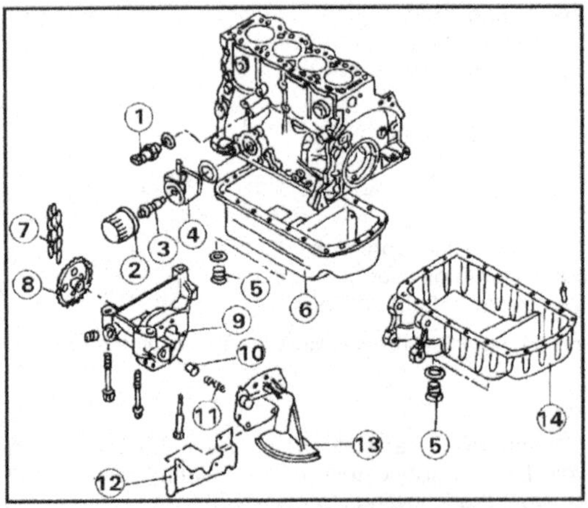

Check the oil pressure piston and the spring (the pump must be dismantled for this purpose). Clean out the piston bore and check for grooves or pitting.

Fig. 1.68 – Attachment of oil pump, oil sump(s), oil filter and other parts.

1 Oil pressure switch
2 Oil filter
3 Bolt for oil filter
4 Oil cooler
5 Oil drain plug and seal
6 Oil sump, steel
7 Pump drive chain
8 Pump drive gear
9 Oil pump
10 Plunger, relief valve
11 Spring, relief valve
12 Oil deflector
13 Oil suction pipe
14 Oil sump, aluminium

A sticking piston or one with excessive clearance can lead to oil pressure problems later on. Replace the piston, spring or the oil pump accordingly.

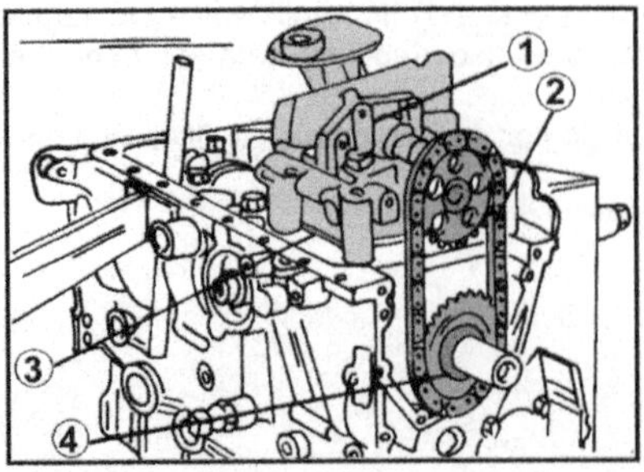

Fig. 1.69 – Fitted oil pump.

1 Oil pump
2 Oil pump drive chain
3 Spacer plate, if fitted
4 Pump drive gear

Before fitting the pump, clean the cylinder block face, without using sharp tools. Place the pump in the correct position, at the same time engaging it with the drive chain, and push it over the dowel sleeve. Tighten the bolts to 0.8 kgm. Refit the oil sump as described during the assembly of the engine. After the installation, refer to Fig. 1.69 although the pump is shown on a removed engine.

To make sure the pump has enough oil when the engine is started, disconnect the plug from the fuel shut-off valve on the HP pump and crank the engine (30 seconds) with the starter motor. Re-connect the cable plug and start the engine. Re-check the oil level after the engine has reached its operating temperature and, if necessary, correct.

1.6.3. OIL FILTER

The oil filter has an integral by-pass valve and only a replacement filter with such a valve must be fitted (Purflux filter LS867A or LS 867B, see Section 1.6.0.). The filter is fitted to the outside of the oil cooler (where the cooling hoses are attached). To remove the filter, use a filter wrench to unscrew the filter, as shown in Fig. 1.70. If none is available, try a piece of emery cloth. Place this around the outside of the filter, with the abrasive side towards the filter body, and with the two hands, try to unscrew the filter.

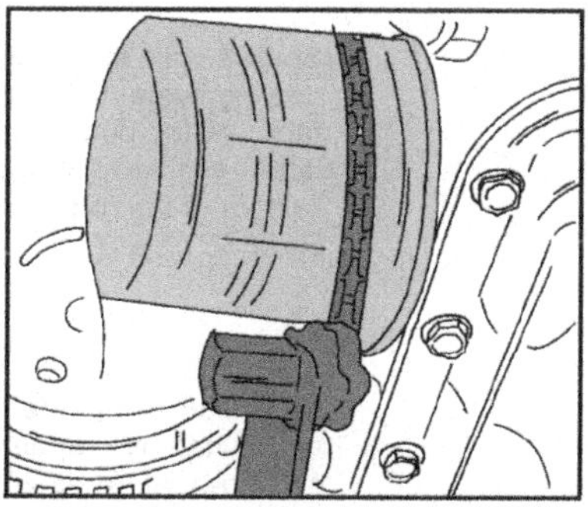

Fig. 1.70 – Removal of the filter with a chain-type filter wrench. Any other filter wrench is, of course, also suitable.

Otherwise, drive the blade of a screwdriver blade through the filter and unscrew it.

Lubricate the seal of the new filter with engine oil and tighten the filter until it touches the cylinder block. From this position, and with the hands only, tighten the filter a further ¼ of a turn. Unscrew the filter once more and re-tighten it until it touches the cylinder block and now tighten it a further ½–¾ of a turn. These instructions must be followed to obtain a perfect seal. The fitting instructions are normally printed on a label, attached to the filter.

1.6.4. OIL COOLER

The oil cooler is fitted between the cylinder block and the oil filter can be seen in Fig. 1.66. As the oil filter must be removed to take out the oil cooler, you will need a filter wrench of the type shown in Fig. 1.70 or similar to remove the filter without damaging it. If there is no reason to replace the filter, it can be refitted. Removal and installation is carried out as follows.

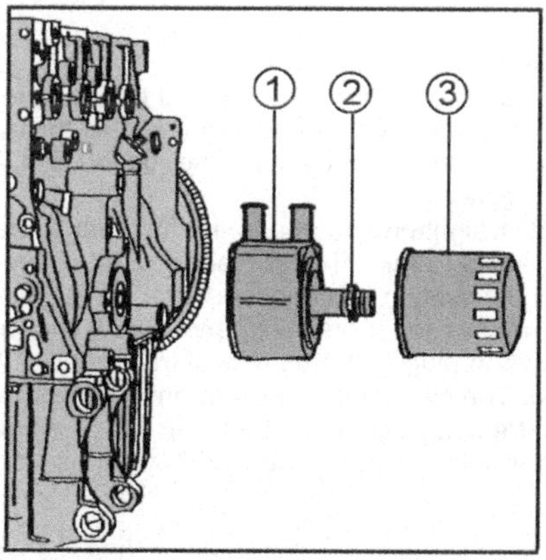

Fig. 1.71 – Arrangement of oil cooler (1) and oil filter (3). The connection (2) is held in position by a bolt and is also used to secure the oil filter.

- Apply the handbrake, jack up the front of the vehicle and place chassis stands underneath the vehicle.
- Drain the cooling system (Section 1.7.) and disconnect the hoses from the oil cooler. It is also possible to clamp the hoses to prevent the coolant from running out, but a slight loss of fluid cannot be prevented.
- Place a suitable container underneath the filter and remove it as described above.

- Remove the bolt in the centre of the oil cooler and remove the cooler.

Refit the oil cooler as follows.

- Place a new oil seal into the groove at the rear of the oil cooler, place the oil cooler in position and tighten the securing bolt.
- Refit the oil filter as described above.
- Lower the vehicle on its wheels and fill the cooling system or add additional coolant to the correct level.
- Check the engine oil level and correct as necessary. Start the engine and check all connections (oil and water) for leaks.

1.7. Cooling System

1.7.0. TECHNICAL DATA

Type of system	Sealed with expansion bottle. Oil cooler connected to cooling system. System consists of radiator, expansion tank, water pump, thermostat and electric fan(s); the latter switched by means of thermo switch at the bottom of the radiator
System capacity	Refer to page 7
Water pump:	
-Type and drive	R.H. side of cylinder block, driven by timing belt from crankshaft. No repair possible
Radiator	Mounted at front of engine, below the lower, inner crossmember
Expansion tank	Plastic container, fitted to R.H. side of engine compartment. Opening pressure of cap 1.0 bar
Anti-Freeze Content:	
-Down to −15 °C	28% anti-freeze
-Down to −30 °C	50% anti-freeze
-Recommended type	Follow manufacturer's recommendation
Changing of anti-freeze	Every 60 000 km or 2 years
Thermostat:	Wax element in separate housing at front of cylinder head. Opening temperature 83 °C, fully open at 110 °C
Electric Cooling Fan:	
-Cutting-in 1st stage	97 °C
-Cutting-in 2nd stage	102 °C

1.7.1. DRAINING AND REFILLING

Two coolant drain plugs are fitted and both should be opened to drain the cooling system. Some radiators may be built without drain plug. A switch is used as a safeguard for overheating. The switch operates when the engine overheats and a warning light in the dashboard "blinks". Investigate the cause immediately, i.e., switch off the engine.

The filling of the system is not straightforward due to the fact that some of the coolant hoses are higher than the upper level of the expansion tank. The workshop uses a special filling bottle, but with some knowledge, you will be able to fill the system as described later on.

To drain the cooling system, remove the protective panel below the engine and unscrew the drain plug in the cylinder block and unscrew the plug at the bottom of the radiator. The expansion tank cap must be removed to facilitate the draining. The system is filled with anti-freeze and this can be collected if wanted. If it is intended to renew the anti-freeze, disconnect the lower radiator hose to drain the system, but catch as much of the anti-freeze as possible – **Avoid Pollution!** – dispose of the anti-freeze in a responsible manner.

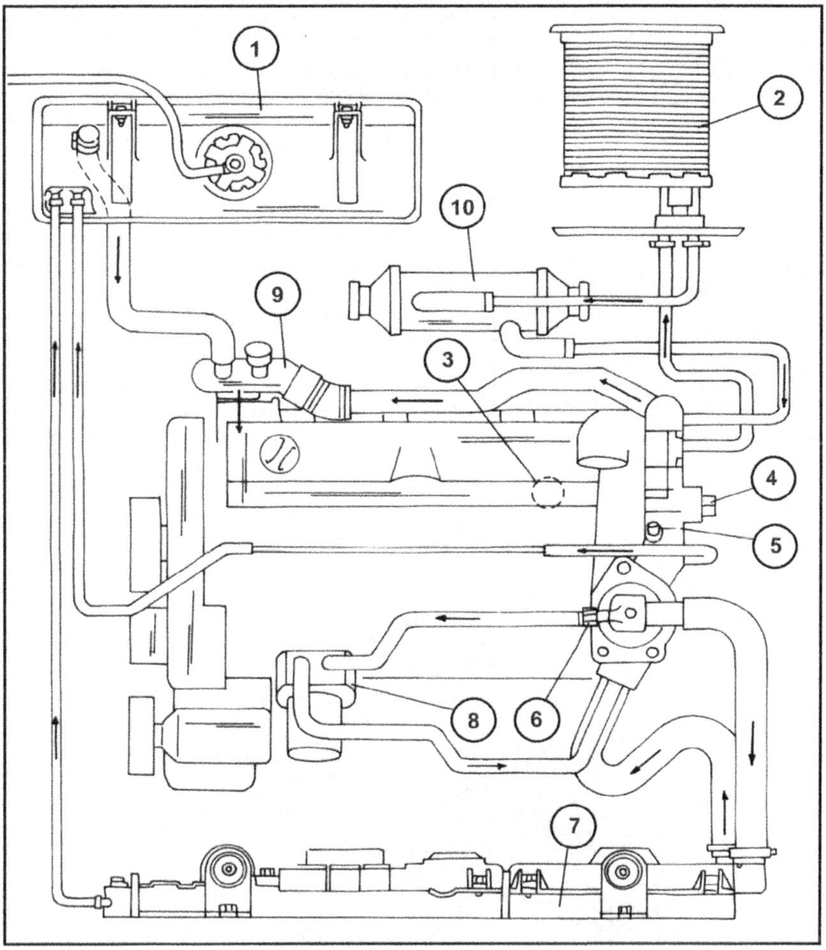

Fig. 1.72 – Cooling system diagram.

1 Expansion tank
2 Heater radiator
3 Drain plug in cylinder block
4 Coolant temperature sensor
5 Coolant outlet housing
6 Bleeder screw
7 Radiator
8 Heat exchanger (oil cooler)
9 Coolant inlet housing
10 Heat exchanger (exhaust gas/water)

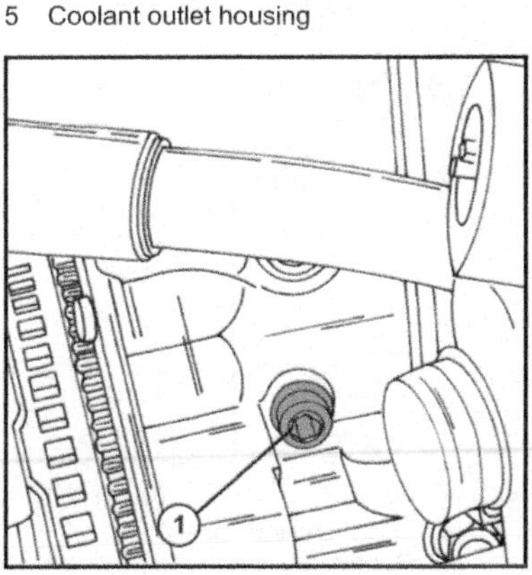

Prepare the anti-freeze mixture in accordance with the temperature you expect in your area. Manufacturer's charts will give you the correct ratio, but remember that only an anti-freeze suitable for these engines must be used. Refer to Section 1.7.0. for general temperatures.

Refit the cylinder block and radiator drain plugs (Fig. 1.73 shows the plug in the cylinder block) and make sure that they are tight. Open the bleeder valve on the radiator (6, Fig. 1.72).

Fig. 1.73 – The location of the drain plug (1) on the cylinder block.

As already mentioned, a filling bottle is used in a workshop. This has the shape shown in Fig. 1.74 and is fitted in place of the expansion tank filler cap. The illustration shows, of course, the special connector (2) and the bottle (1) belonging to the special tool set. However, you may use a plastic bottle that you cut on both ends. Prepare water tight joint between the bottle end and the filler opening and fill the system. You will need a helper to hold the bottle in place when you fill the system.

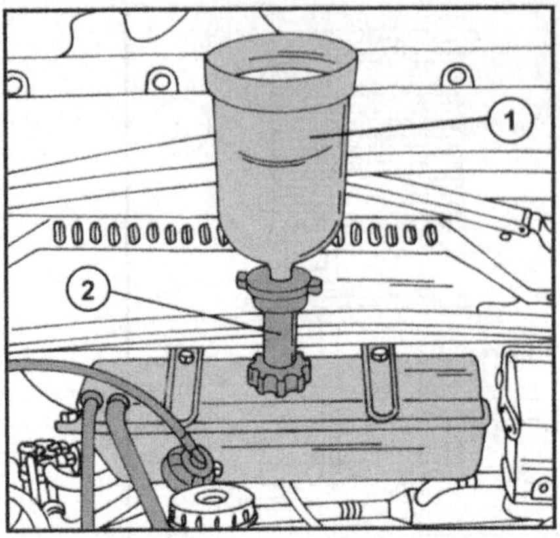

Fig. 1.74 – Using the filler bottle (1). The bottle is joint with a connection (2) to the expansion tank opening.

Fill the coolant through the expansion tank opening and close the bleed screw as soon as only liquid, free of air bubbles, comes out of the bleeder screw hole and the coolant level is up to the "Max" mark on the expansion tank. Check that the filler bottle is still full to the rim.

Start the engine (*do not run the engine with the bleeder screw slackened)* and run it at 1500–2000 rpm. Wait until the thermostat has opened or the cooling fans have cut-in and then allow the engine to run approx. 1 minute at idle speed.

Remove the filling bottle, but take care, as the coolant in bottle can also be hot. The surplus coolant in the bottle will drain off, but the system is filled.

Switch off the engine and fit the expansion tank cap. Allow the engine to cool down for at least 3 hours and re-check the coolant level. The coolant must be level with the "Max" mark. Otherwise correct. Take the opportunity to check the system for leaks.

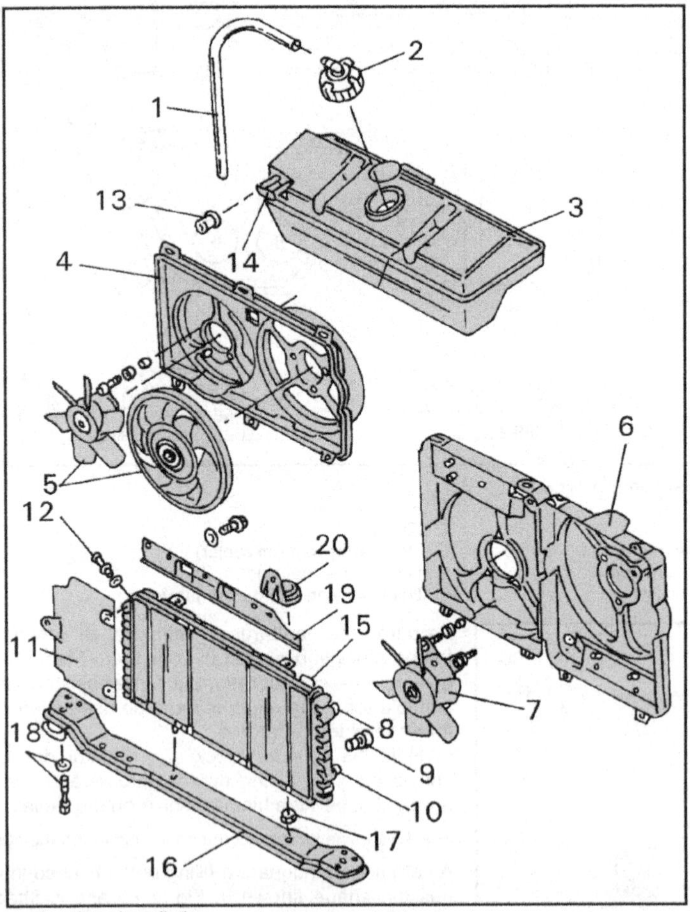

Fig. 1.75 – Radiator and radiator installation.

1 Overflow hose
2 Expansion tank cap
3 Expansion tank
4 Radiator/shroud (A/C)
5 Fan assembly, with A/C
6 Radiator shroud
7 Fan assembly without A/C
8 Elbow, upper hose
9 Thermo switch, fan
10 Elbow, lower hose
11 Air guide panel
12 Connection to tank
13 Connection to radiator
14 Connection to cyl. head
15 Radiator
16 Lower crossmember
17 Rubber grommet
18 Bolt, spring washer
19 Upper crossmember
20 Upper securing bracket (radiator)

1.7.2. RADIATOR – REMOVAL AND INSTALLATION

The removal and installation of the radiator is identical with the removal of the front panel of the engine, described during the removal of the engine. Smear the thread of the temperature switch with sealing compound (if removed) and screw the switch into the radiator. If the radiator has been replaced, transfer the rubber mountings to the new radiator. After installation, fill the cooling system as described above. Fig. 1.75 shows the radiator together with the associated parts.

1.7.3. WATER PUMP – REMOVAL AND INSTALLATION

The water pump cannot be repaired or serviced. If the pump is faulty or leaking, fit a new unit.

The pump is located on the timing side of the engine and is driven by the timing belt, i.e., it is hidden out of view. The pump can be removed with the engine fitted to the vehicle. The protection panel inside the wheel arch must be removed. Remove and refit the timing belt as described to gain access to the water pump. Fig. 1.56 shows with (4) where the water pump is located. Unscrew the water pump from the cylinder block.

1.7.4. THERMOSTAT

The thermostat is located inside the water elbow on the cylinder head, below a cover (where the hose is connected). To remove the thermostat, drain the cooling system (Section 1.7.1.) and disconnect the hoses from the water outlet. One hose leads to the oil cooler, and the other one leads to the lower radiator connection. Unscrew the cover from the thermostat housing and lift out the thermostat. A rubber sealing ring is inserted between cover and housing.

Fig. 1.76 – Checking a thermostat.

A thermostat cannot be repaired or adjusted and must be replaced if faulty. A thermostat can be tested by immersing it in a container of cool water and gradually raising the temperature to check that it opens properly. Suspend the thermostat on a piece of wire so that it does not touch the sides or the bottom of the container. A thermometer must be suspended in the same manner. The arrangement is shown in Fig. 1.76. The opening temperature is marked in the case and can be used as a guide during the test. The opening temperature and the "fully open" temperature are given in the Technical Data. A difference of 1 °C is permissible in each case.

Use a new sealing ring when fitting the thermostat. Check the condition of the hose or hoses before connecting. Hoses are sometimes cut at the positions of the hose clips. In this connection, also check the hose clips. Refill the cooling system.

If a new thermostat is fitted, take the vehicle on a run and check that the thermostat opens and closes as designed.

Thermostat – Trouble Shooting

Thermostat does not open. It may stick or one of the valve plates does not release. The temperature warning is in the red zone and the warning light lights up, but the radiator is not hot. Wait until the engine has cooled off and then remove the thermostat and take it out. Driving without thermostat for a short while will not harm your engine, driving with a closed thermostat could lead to extensive engine damage.

Thermostat does not close properly. Foreign matter could have entered. In this case, the radiator will be fairly hot not long after the engine has been started. The heater system will take long to supply warm air. The thermostat must be replaced as soon as possible to prevent damage to the engine.

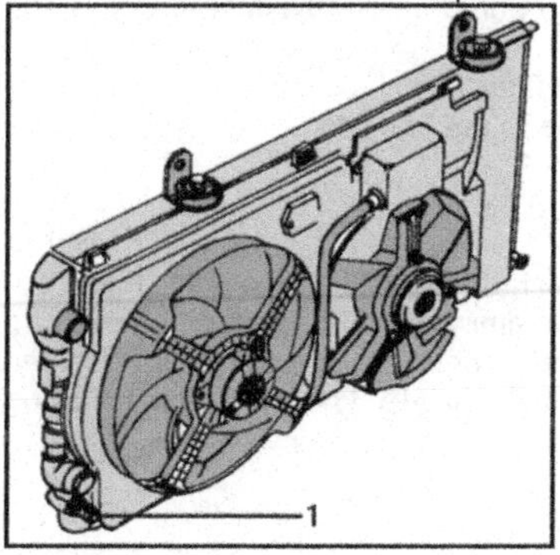

1.7.5. ANTI-FREEZE SOLUTION

The cooling system is filled with anti-freeze and the mixture should be left in the system throughout the year. If a fresh mixture is prepared, use anti-freeze and mix with water (50% anti-freeze; 50% water will protect the system down to −30 °C). We advise you only to use the anti-freeze marketed by the relevant manufacturer, as the solution has been specially formulated for use with your engine. If the system is topped-up with plain water, remember that the anti-freeze solution will be diluted. It is always best to mix anti-freeze solution with water, even when topping up.

Fig. 1.77 – The radiator is removed from the vehicle as shown and can then be dismantled. The thermo switch (1) regulates the fan operation.

1.7.6. COOLING SYSTEM TEMPERATURE SWITCHES

Fan Temperature Switch: The switch is fitted to the rear of the lower radiator part, as shown in Fig. 1.77. The switch has the function to switch the cooling fans on and off in accordance with the coolant temperature.

To replace the switch, drain the cooling system, withdraw the cable connector plug and unscrew the switch. The installation is a reversal of the removal procedure. Make sure that the sealing washer is in good condition. Refill the cooling system and check the proper operation of the fans.

Coolant Temperature Switches: These switches are fitted into the thermostat housing. Removal and installation is straightforward. Check the sealing washer(s) when fitting the switch(es). The switches are tightened to 1.8 kgm (13 ft.lb.).

1.8. HDi FUEL INJECTION SYSTEM

The following information will help you to familiarise yourself with the fuel injection system with high-pressure common rail injection, made by Bosch, with the Ref. No. EDC 15C7. Fig. 1.78 shows the parts used in the system.

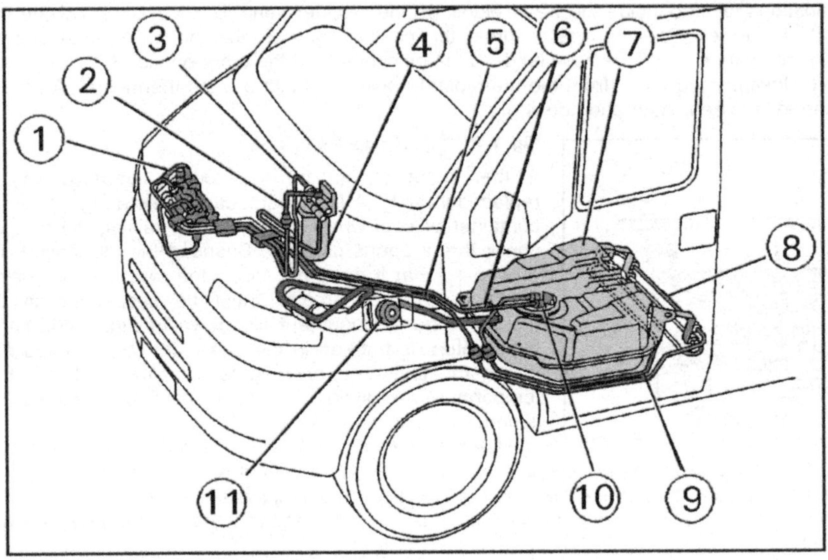

Fig. 1.78 – The various parts of the fuel system on the HDi engine. Refer to the text.

- The high-pressure fuel pump (1) is located on the R.H. of the vehicle.
- The return pipe (2) is fitted between the fuel filter and the low-pressure return connection.
- The pipe (3) is fitted between the fuel filter to the high-pressure pump.
- The fuel filter (4) is connected with various pipes between the high-pressure fuel pump and the tank.
- The pipe (5) is fitted between the low-pressure return connection and the fuel cooler.
- The pipe (6) is connected between the fuel tank and the fuel filter.
- The fuel tank (7).
- The fuel cooler (8) cools the returned fuel from the high-pressure pump before the fuel re-enters the tank (not fitted with the 2-litre engine).
- The pipe (9) is fitted between the fuel cooler and the fuel tank (not fitted).
- The booster pump delivers the fuel under low pressure via the pipe to the fuel filter.
- The fuel tank filler cap.

High-Pressure Fuel Pump

The high-pressure fuel pump, as shown in Fig. 1.79, is made by Bosch and is of the type CP1. It is a so-called Radialjet Pump with 3 radial pistons. The pump is driven by the timing belt. There are, however, no settings of the drive gear to be observed, when the timing belt has been removed. The pump is lubricated and cooled by the fuel. The max. flow pressure of the pump is 1350 bar (19,170 psi.). The following information refers to Fig. 1.79.

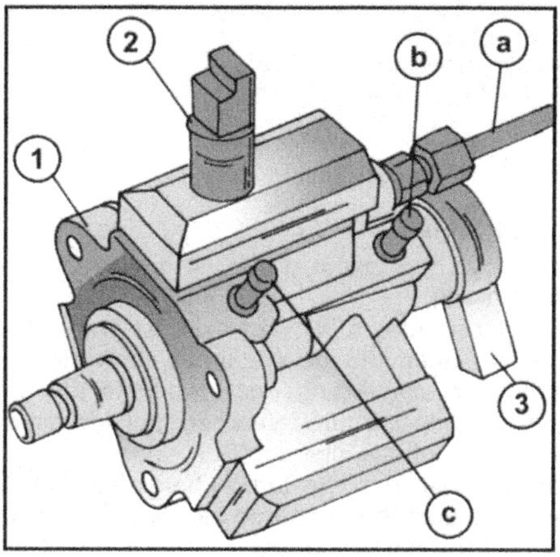

Fig. 1.79 – View of the high-pressure fuel pump. The numbers and letters are referred to the description.

- The high-pressure pump (1) is fitted with a so-called deactivator (2) for the 3rd piston of the pump.
- A high-pressure fuel regulator (3) is fitted at the position shown to control the pressure within the pump and the system.
- The pipe connection (a) is the fuel outlet towards the common injection rail, as shown in Fig. 1.80. Excessive fuel is returned via connection (b) to the fuel tank but enters the fuel cooler before doing so. The fuel enters the pump via connection (c).

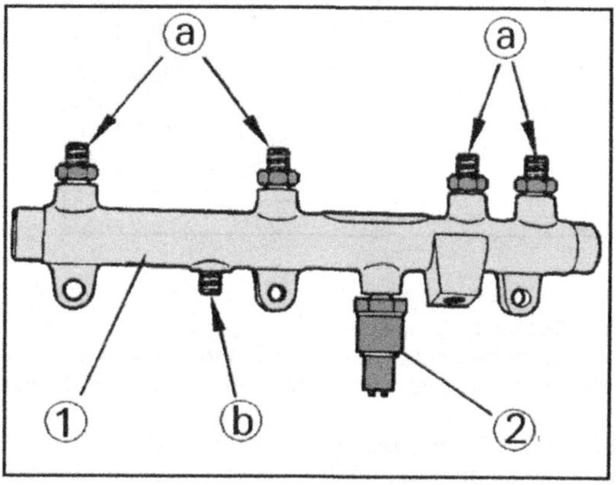

Fig. 1.80 – View of the common injection rail (1). Fitted to the bottom is the high-pressure regulator. The connections (a) are connected with the diesel fuel injectors, and the fuel enters the rail via connector (b).

High-Pressure Common Injection Rail

In a similar manner as known from petrol fuel injection systems, a single rail (common rail) distributor is used to direct the fuel to the four injectors. The high-pressure fuel pressure regulator is fitted to the pump, as shown in Fig. 1.80.

The fuel pressure regulator controls the pressure of the fuel at the outlet of the high-pressure fuel pump.

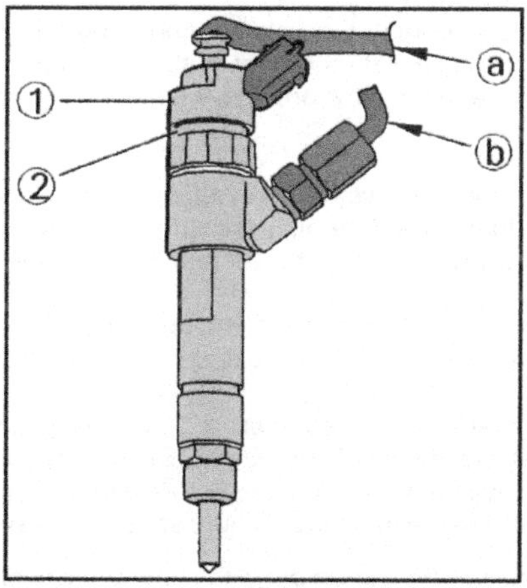

Fig. 1.81 – View of an injector as used with the common rail fuel injection. Numbers and letters are referred to below.

Diesel Fuel Injectors

The fuel injectors of this system are different in construction, when compared with conventional injectors. They are made by Bosch (96 255 425 80). A cable connector (1) is fitted to each injector to operate the injection control electro-valve (2). Connection (a) is connected over pipes to the fuel tank, i.e., excess fuel is returned via the fuel cooler. The fuel under pressure from the injector rail enters each injector via connection (b), secured to the injector by a union nut. The connection must not be removed from the injector.

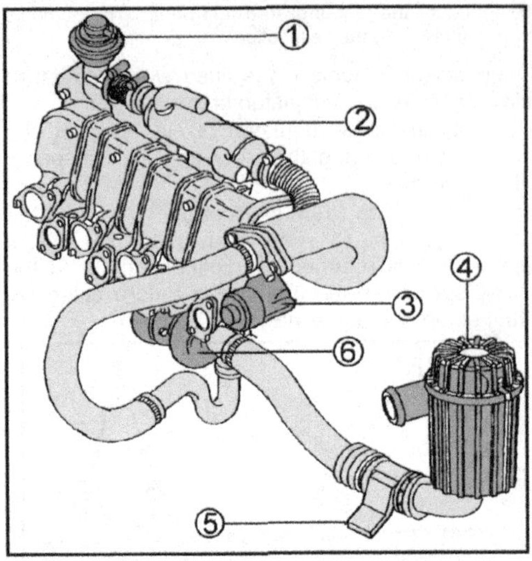

Fig. 1.82 – Layout of the air supply circuit.
1 EGR valve (electric)
2 Coolant/re-cycled gas exchanger
3 Throttle valve housing
4 Air cleaner
5 Air flow meter
6 Turbo charger

Air Supply Circuit

Fig. 1.82 shows the layout of the air supply circuit. Note the following.

- The engine operates with an exhaust gas turbo charger (6), fitted into the system as shown in the illustration. A connecting duct connects the air cleaner (4) with the turbo charger. Outside air is drawn in through the inlet duct.
- Excessive heat from the turbo charger is routed through the duct to the charge air cooler (2), also referred to as heat exchanger.
- The cooled air is directed through the heat exchanger to the inlet throttle housing (3).
- The turbo charger is mounted directly on the exhaust manifold.

Fuel Delivery Pump

The fuel delivery pump is located in the tank and operates electrically with the delivery pressure of 2.5 bar (35.5 psi.). The pump has a delivery capacity of 155 litres per hour at a voltage of approx. 13.5 volts.

Additional Component Parts of the System

An accelerator pedal potentiometer is fitted to the pedal. The position of the pedal is converted into an electrical signal, which is transmitted to the electronic control unit to calculate the injection times and fuel pressure for the best operation of the engine. The pedal travel can be adjusted, but a diagnostic instrument is required.

Brake Pedal and Clutch Pedal Switches

The brake pedal has a switch that controls the vehicle stop lamps. From the "brake pedal depressed" signal, the electronic control unit will calculate the deceleration situation. The clutch pedal switch enables the control unit to differentiate between "in gear" and "gear changes".

Repairs on the System

Because of the high-pressure operating system, we refrain from giving your advice on removal and installation instructions for individual parts. The high-pressure pump can be removed after the timing belt has been removed. At all times, make sure that the system is not under pressure, when connections are opened.

If, however, intent to carry out any work on the system, note the following precautions.

- Never work on the system with naked flames, and under all circumstances, prevent creation of sparks through short circuits.
- Always switch off the engine at the slightest sign of fuel leakage. Under no circumstances, try to prevent to stop a spray of fuel with the tip of your finger. Switch off.
- After switching off the engine, wait at least 1 minute before carrying out any work. The system needs a certain time to bring the high-pressure fuel back to atmospheric pressure.

Some of the operations described can be carried out with some precautions.

High-Pressure Injection Pump – Removal and Installation

A puller to remove the pump drive gear will be required to remove the pump. Make sure that the system is not under pressure.

- Disconnect the battery earth cable.
- Remove the drive belt at the front of the engine and remove the timing belt as already described. The R.H. engine mounting removed during the operations must be refitted.

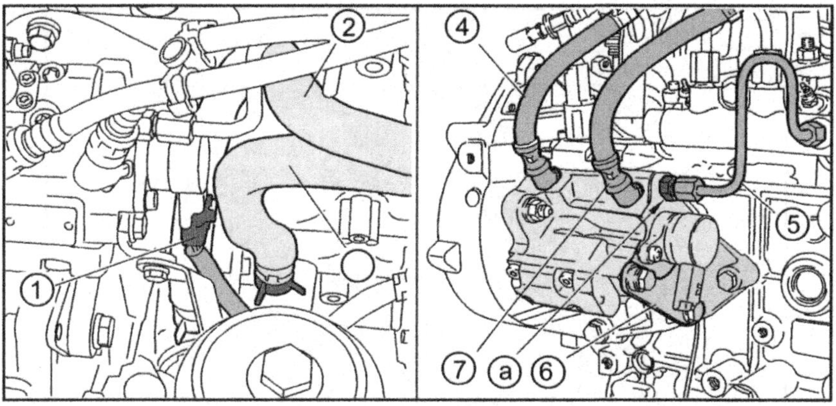

Fig. 1.82a – Details for the removal and installation of the high-pressure pump. The numbers are referred to the text.

- Refer to Fig. 1.82a and remove the connector (1) and the two hoses (2) and (3). Remove the high-pressure pipe (5) and the rear attachment of the pump (6). Two open-ended spanners must be applied at position "a" when the pipe (5) is removed from the pump.

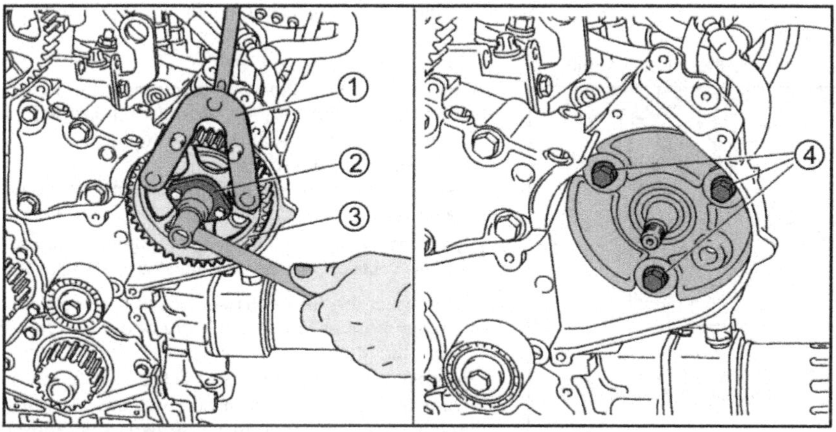

Fig. 1.83 – Removal of the high-pressure pump. On the L.H. side, the pump drive gear is removed. The R.H. view shows the attachment of the pump to the engine.

- Disconnect the fuel feed pipe (4) and the fuel return pipe (7). All openings in the pump must be closed off in a suitable manner to prevent the entry of dirt.
- The puller (2) must now be applied as shown in Fig. 1.83. Remove the nut securing the pump drive gear and fit the puller with the two bolts. Use a suitable socket and remove the drive gear (3). A suitable holder is used to prevent the drive gear from rotating.
- With the drive gear removed, you will see the three bolts securing the high-pressure pump. Remove the bolts and withdraw the pump.

The installation is a reversal of the removal procedure. The high-pressure pipe (5) in Fig.1.82a must always be replaced. Replace the timing belt as described earlier on. Observe the following tightening torques.

- The bolts (4) in Fig. 1.83 are tightened to 2.0 kgm (14.5 ft.lb.).
- The rear attachment (6) in Fig. 1.82a is tightened to 2.25 kgm (15 ft.lb.).
- Use open-ended spanners and tighten the high-pressure pipe (5) in Fig. 1.82a to 2.5 kgm (18 ft.lb.).
- Fit the drive gear and the nut, prevent the gearwheel from rotating, as shown in the L.H. view of Fig. 1.83, and tighten the nut to 5.0 kgm (36 ft.lb.).

After installation, start the engine and check for leaks on the connection of the high-pressure pipe (5).

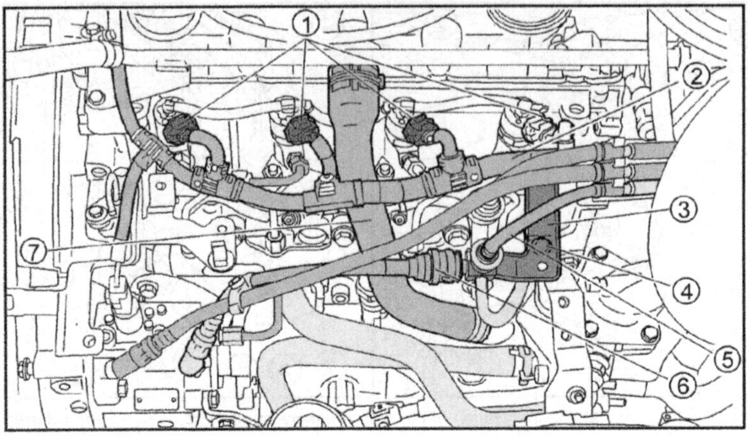

Fig. 1.84 – Details for the removal and installation of the common injection rail. The numbers are referred to the text.

High-Pressure Common Injection Rail – Removal and Installation

Needless to say that the operation must be carried out under clean condition. Refer to Fig. 1.84 and proceed as follows.

- Disconnect the battery earth cable.
- Refer to the illustration and disconnect the four connectors (1) at the positions shown. The wiring harness support bar (2) must be moved to one side in order not to interfere with the following operations.
- Disconnect the hose (7) and move it out of the way. This is the oil vapour re-circulation hose for the engine.

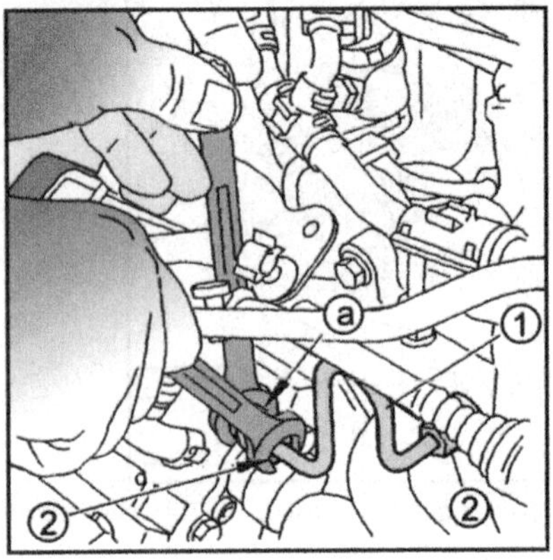

Fig. 1.85 – Disconnecting the high-pressure fuel pipe (1). To slacken the two union nuts (2), apply an open-ended spanner as shown by "a".

- Disconnect the tubes (5) and (6) and close their ends in a suitable manner to prevent entry of dirt and free them from their attachment.
- Remove a bolt (4) and push the bracket (3) to one side.
- Refer to Fig. 1.85 and disconnect the high-pressure fuel supply pipe. To disconnect, use two open-ended spanners as shown in the illustration and slacken the two connectors at the ends of the pipe. Make sure no dirt can enter the connections. Remove the pipe.

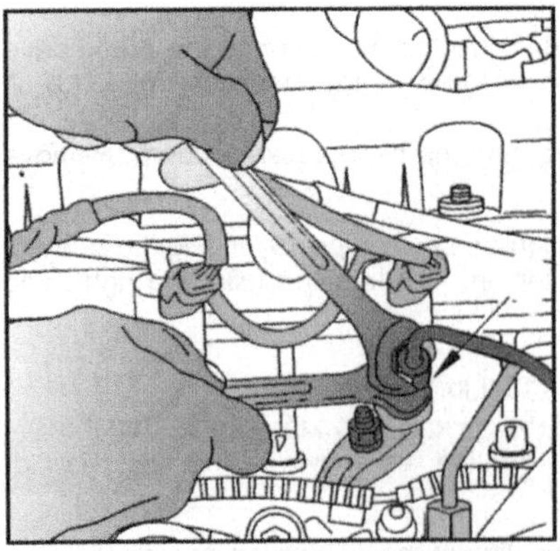

Fig. 1.86 – Disconnecting the high-pressure fuel pipe from the injectors. The two open-ended spanners are applied as shown to remove the pipe union nut (arrow).

- In a similar manner, remove the high-pressure pipes from the diesel injectors as shown in Fig. 1.86. Again, two open-ended spanners are used. Slacken the union nuts at the injector end and the unions at the common rail ends and then remove the high-pressure pipes. The open connections must be plugged in to prevent the entry of foreign manner.
- You will now have the view shown in Fig. 1.87. At the bottom of the common rail, unscrew the high-pressure sensor (3). Then, remove the bolts (1) and take out the common rail (2).

The installation is a reversal of the removal procedure. High-pressure injection pipes must always be replaced. Note the following points.

- Fit the common rail (2) in Fig. 1.87 and fit the screws (1) and tighten finger-tight. Fit the high-pressure pipe (1) in Fig. 1.85. The two unions (2) are tightened to 2.2 kgm (15 ft.lb.). Again, two open-ended spanners are used to tighten the union nuts. You will, therefore, estimate the torque as no torque wrench can be applied.

- Refit the diesel injector high-pressure pipes to the injectors of cylinders No. 2 and No. 4 and secure the pipes with the union nuts finger-tight on the injectors and the common rail. Now fit the high-pressure pipes to the cylinders No. 1 and No. 3 in the same manner. Make sure that the pipe clips are in their original position.

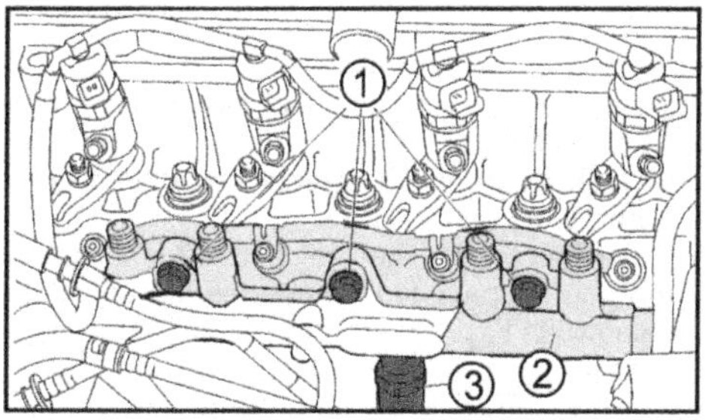

Fig. 1.87 – The bolts (1) secure the common rail (2). At the bottom, you will find the pressure sensor (3).

- Tighten the high-pressure union nuts to 2.5 kgm (18 ft.lb.). Again, two open-ended spanners are used and the torque will have to be estimated unless a claw-type wrench is available as used in a workshop. The pipes are tightened to the same torque on the common rail side and on the diesel injector side.
- Tighten the bolts (1) in Fig. 1.87 to 2.5 kgm (18 ft.lb.).
- All other operations are carried out in reverse order to the removal procedure.

Diesel Fuel Injectors – Removal and Installation

A long socket is required to remove a diesel fuel injector (a special socket is used in the workshop). We must point out that it is rather difficult to reach some of the union nuts. We advise you, therefore, to first check the area before you proceed with the work.

- Disconnect the battery earth cable.
- Refer to Fig. 1.84 and withdraw the four plugs (1) from the injectors. The wiring harness support bar (2) must be moved to one side in order not to interfere with the following operations.

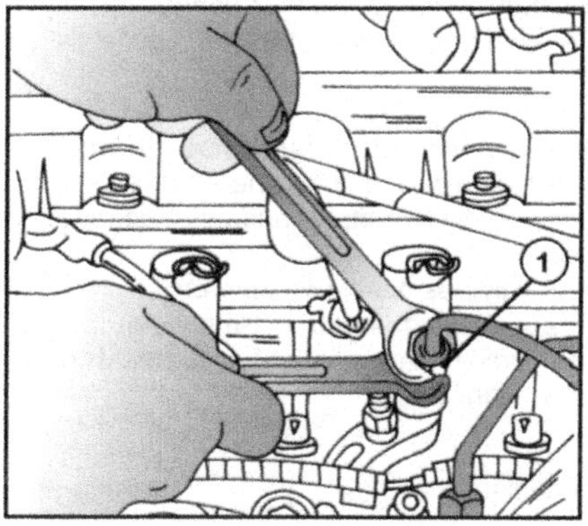

Fig. 1.88 – When slackening the union nut (1) of the high-pressure pipe on injector and common rail, counterhold the hexagon with an open-ended spanner and use the second spanner to undo the union nut.

- Disconnect the fuel return pipes that are connected to the injectors.
- Looking at the engine compartment, unscrew and remove the support, situated below the L.H. connector (1) in Fig. 1.84.
- Clean the high-pressure unions before disconnecting them. This is important. For example, in a workshop, the mechanics are advised to wear clean overalls to disconnect the unions.
- Apply two open-ended spanners in the manner shown in Fig. 1.88 to the pipe union nut and unscrew the pipe from the injector. In the same manner, disconnect the pipe from the union on the common rail. Plug the openings on the injector and the common rail in a suitable manner. The workshop uses special caps that are pushed over the open connections.

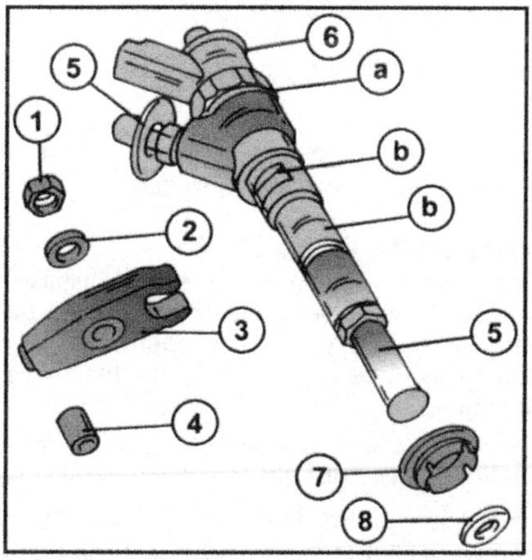

Fig. 1.89 – View of a diesel injector. Never slacken the nut "a". If it is difficult to remove, an injector follows the instructions below.

1 Nut, 3.0 kgm
2 Cup
3 Injector securing flange
4 Distance sleeve
5 Protective cap (special tool)
6 Diesel injector
7 Seal
8 Seal

- Remove the nut (1) in Fig. 1.89, take off the cap (2), i.e., a washer, and lift off the flange (3). At the bottom of the injector, remove the seals (7) and (8). Sometimes, the injector is difficult to remove. Should this be the case, remove the securing stud for the flange and apply an open-ended spanner at the flat "b" in Fig. 1.89. Use the spanner as a lever to remove the injector.

The installation of the injector is carried out as follows.

- Remove the caps used to close off the open connections and place the distance sleeve (4) in position. Fit the two seals (7) and (8). Both must be replaced.
- Insert the injector (5), place the securing flange (3) in position and place the washer (2) over the flange. Fit the nut (1) without tightening it at this stage.

The high-pressure injection pipes must be fitted in a certain order. Proceed as follows.

- Fit the injection pipes for cylinder numbers 2 and 4 finger-tight to the injectors and the common rail. Then, tighten the nuts (1) in Fig. 1.89 of the two cylinders to 3.0 kgm (22 ft.lb.).
- Fit the injection pipe for cylinder numbers 1 and 3 finger-tight to the injectors and the common rail. Then, tighten the nuts (1) securing the flanges (3) to 3.0 kgm (22 ft.lb.).
- Tighten the union nuts on the injectors and on the common rail to 2.5 kgm (18 ft.lb.). As a special wrench is used in the workshop to apply a torque spanner, you will have to estimate the torque value. The hexagon below the union nuts must be held firmly as shown in Fig. 1.88 to prevent it from moving.
- All other operations are carried out in reverse order to removal.

Turbo Charger – Removal and Installation

Fig. 1.90 shows the details of the removal and installation. The turbo charger is removed together with the exhaust manifold.

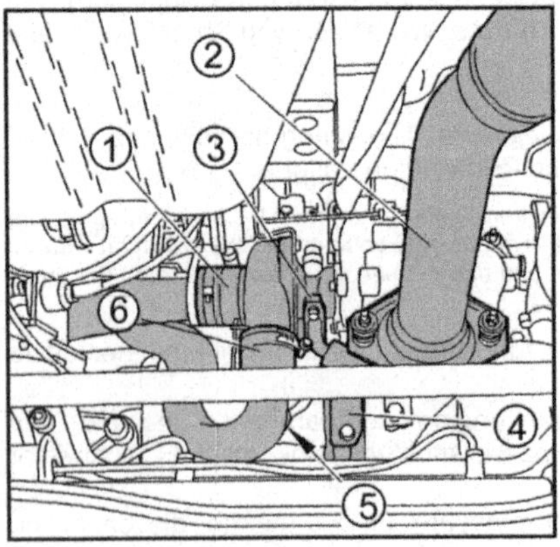

Fig. 1.90 – Details for the removal and installation of the turbocharger. The numbers are referred to the text.

- Disconnect the negative battery cable and jack up the front end of the vehicle. Remove the protective panel underneath the engine compartment.
- Disconnect the exhaust pipe (2) from its flange.
- Remove the turbo charger inlet duct (1) and the outlet flange (6).
- Unscrew the support bracket (4).
- Disconnect the turbo charger oil feed pipe (5) and the turbo charger oil return pipe (3) on the engine and the turbo charger side. The open pipe ends must be plugged in a suitable manner.
- Remove the EGR valve as described later on.
- Remove the nuts securing the exhaust manifold, take off the spacers and remove the manifold together with the turbo charger. The turbo charger can now be removed from the manifold.

Carry out the following checks before installation.

- Check the air cleaner system and all connections for good condition.
- Change the engine oil and the oil filter.
- Make sure that the oil circuit and the union nuts are clean. Use new sealing rings where necessary. Use only the bolts specified for the turbo charger, as these must be heat-resistant.
- Make sure that the air intake hoses and the exhaust manifold are free of foreign matter.

Install as follows.

- Fit the turbo charger to the exhaust manifold and attach the assembly to the engine.
- Refit the removed parts in reverse order to removal. Finally, re-connect the battery earth cable.

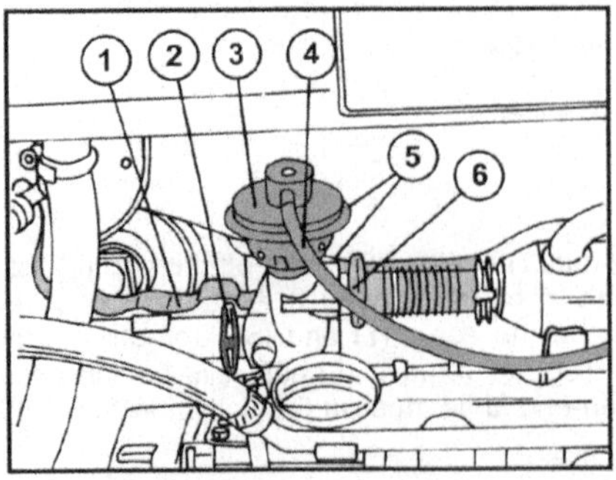

Exhaust Gas Re-Circulating Valve

The valve can be replaced, but no repairs are possible. Fig. 1.91 shows where the valve is located. It should be noticed that there is also an electrically operating EGR valve. The following instructions only refer to the valve shown in the illustration.

Fig. 1.91 – Details for the removal and installation of the EGR valve. The numbers are referred to the text.

- Disconnect the battery earth cable.
- Withdraw the heat shield (1) and disconnect the hose (4) from the EGR valve (3).
- Slacken the hose clamps (2) and (6). The clamps are

spring clamps and normally a special pair of pliers suitable for spring clamps is used.

- Remove the bolts (5) located behind the valve and take out the valve together with the seal.

The installation is a reversal of the removal procedure.

1.8.1. FUEL FILTER

The fuel filter is shown in Fig. 1.92. The filter is fitted with a replaceable filter cartridge. A thermostatic sensor (3) controls the fuel temperature. If the temperature of the fuel is below 6 °C, an electrically operated heating element heats up the fuel until a temperature of 15 °C is obtained, before it reaches the high-pressure pump. A sensor at the bottom of the filter (5) monitors the water content of the fuel. If water is detected in the filter, a warning light will light up in the instrument panel.

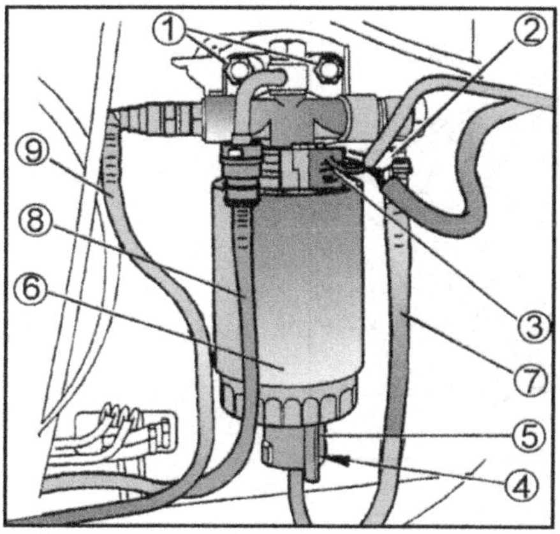

Fig. 1.92 – The fitted fuel filter. The numbers are referred to the description of the filter removal.

The filter is replaced in a similar manner as an oil filter. A new filter cartridge is supplied together with a seal, which is placed between the cartridge and the filter head and must always be used. The following instructions give the necessary operations to remove the complete filter. The cartridge can then be replaced on a bench.

The fuel filter should be replaced every 20 000 miles but must be drained every 6000 miles to empty water collected through condensation.

- Disconnect the battery earth cable and remove the air cleaner.
- Place a small container underneath the filter to catch the draining diesel fuel.
- Withdraw the connector plugs from connectors (2), (3) and (4).
- Disconnect the pipes (7), (8) and (9) from their connections. Protect the open pipe ends in a suitable manner to prevent the entry of foreign matter.
- Remove the nuts (1) and take off the complete filter assembly. At the bottom of the filter, unscrew the water detector (5).
- Remove the filter cartridge on a workbench, using a chain-type filter wrench, in a similar manner as you would remove an oil filter.
- The installation is a reversal of the removal procedure. Place the new seal over the threaded insert and tighten the filter cartridge until it rests against the filter head. From this position, tighten the cartridge by a further one-quarter of a turn.

All other operations are carried out in reverse order to the removal procedure. After installation, rotate the engine with the starter motor for approx. 5 seconds. Repeat the operation until the engine starts.

1.8.2. DRAINING THE WATER IN THE FILTER

Note: The water should be drained from the fuel filter every time the engine oil is changed. After draining, bleed the system as described above.

1.8.3. AIR CLEANER

The air cleaner is fitted with a dry paper element and is fitted to the L.H. side of the inner wing panel. Different installations are used with different engines, depending on the model year and the engine. The air cleaner element should be changed every 40 000 miles in the case of a naturally aspirating engine or every 20 000 miles in the case of a turbo charged engine. Again, these are general instructions and only your service booklet will give you the correct interval to change the air cleaner element.

The replacement of the filter element is straightforward. Make sure that the correct filter element is obtained and fitted.

Never attempt to clean a soiled filter element. Slight dust accumulations can, however, be removed with compressed air, from the inside towards the outside.

1.8.4. PRE AND POST HEATING

The pre-heating time depends on the temperature of the engine. The pre-heating unit is controlled by the electronic control unit for the engine operation. Two suppliers (Nagares and Cartier) are used. Champion CH170 heater plus are used.

2. CLUTCH

The operation of the clutch is not the same on all models, as it depends on the gearbox in the vehicle. The clutch of a vehicle with MG5TE gearbox, the clutch release lever, is "pulled", and with an ME5TU gearbox, it is "pushed". The differences between the two types can be seen in Figs. 2.1 and 2.2. All models are covered in this section. The MG5TU transmission is only used together with the 2.5- and 2.8-litre engines.

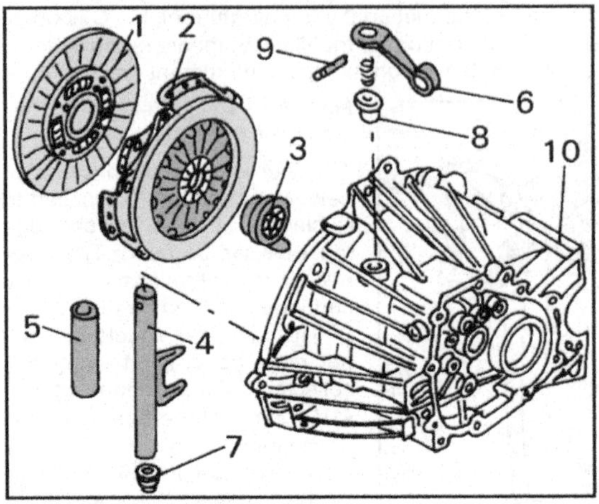

Fig. 2.1 – View of the clutch unit (ME5TU and ML5TU gearboxes, "push" type release.

1 Driven plate
2 Clutch pressure plate
3 Clutch release bearing
4 Clutch release shaft
5 Bearing bush
6 Clutch release lever
7 Release shaft bush
8 Release shaft bush
9 Clamp bolt
10 Gearbox

2.0. Technical Data

Type	Single plate, dry clutch with diaphragm spring.
Clutch operation	By clutch cable
Pedal adjustment	At release lever
Clutch release bearing	Sealed ball-type bearing in constant contact with the clutch plate. "Pulled" or "pushed" as mentioned above.
Clutch Diameter:	228.6 mm

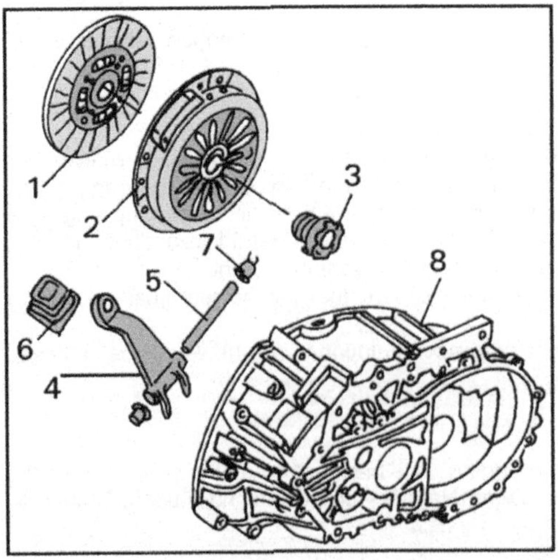

Fig. 2.2 – View of the clutch unit (MG5TU gearbox, "pull" type release) – 2.5- and 2.8-litre engines.

1 Clutch-driven plate
2 Clutch pressure plate
3 Clutch release bearing
4 Clutch release lever
5 Clutch release shaft
6 Rubber dust cap
7 Release shaft bushes
8 Gearbox

2.1. Clutch Unit

2.1.0. CHECKING THE CLUTCH OPERATION

The clutch can be checked for proper operation when fitted to the vehicle. To do this, proceed as follows.

- Start the engine and allow to idle. Depress the clutch pedal and wait approx. 3 seconds. Engage the reverse gear. If grating noises can be heard from the transmission, it can be assumed the clutch or driven plate needs replacement, as the driven plate no longer connects the clutch pressure plate with the flywheel.

- To check the clutch for signs of slipping, drive the vehicle until the clutch and transmission have reached operating temperature. Stop the vehicle, firmly apply the handbrake and engage the 3rd gear. Keep the clutch pedal fully depressed and accelerate the engine to approx. 3000–4000 rpm. Release the clutch pedal suddenly. The clutch operates satisfactorily if the engine stalls immediately.

2.1.1. REMOVAL AND INSTALLATION

Pressure plate and the diaphragm spring cannot be dismantled. Clutches are sometimes changed without prior notice. Only by quoting the model year and the type of engine fitted, you will obtain the correct clutch.

The gearbox or the engine and the gearbox must be removed to replace the clutch plate and/or the driven plate. The following description refers to the removal when the gearbox has been removed, which is generally the case. If the complete power unit engine/transmission has been removed, you will have to separate the transmission from the engine. Remove the cover plate underneath the transmission, remove the TDC sensor from the flywheel housing (bottom of transmission) and remove the starter motor. Suspend the transmission on a lifting tackle and remove the bolts securing the transmission to the engine.

If the transmission has been removed.

- Mark the clutch cover and the flywheel to ensure correct re-assembly, if the parts are refitted. Remove the clutch bolts carefully. The engine must be prevented from rotating when the bolts are slackened. Otherwise, use a ring spanner and place it over the bolt heads and hit the spanner with a short sharp blow to loosen the bolts.
- Remove the clutch. If an MG5TE gearbox is fitted, you will find the clutch release bearing attached to the clutch plate, as shown in Fig. 2.3. Remove the driven plate. Do not allow grease or oil to contaminate the lining faces or any other part, if there is a chance that the parts can be re-used. Make a not which way the driven plate is fitted.

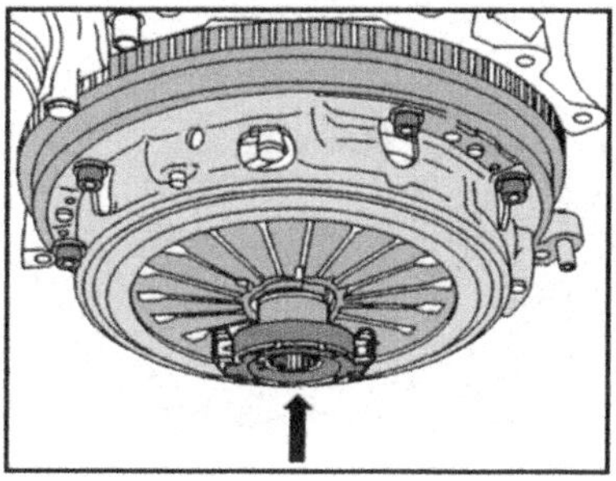

Fig. 2.3 – The clutch release bearing is fitted to the clutch diaphragm spring as shown.

- Clean the inside of the flywheel and check the friction face for wear. If the clutch linings have worn down to the height of the rivet heads, there is a possibility that the rivet heads have scored the flywheel. Replace the flywheel if necessary, but remember that new flywheel bolts must be used. Flywheel can be re-grounded to a certain dimension, but only your dealer will be able to advise you.

The installation is a reversal of the removal procedure. The installation requires normally the use of an alignment mandrel, which is inserted as shown in Fig. 2.4. If this is not available, a spare clutch shaft can also be used. If neither of the above is available and tools cannot be obtained, it is possible to line-up the clutch disc in the inside of the flywheel by inserting a drift of the diameter of the spline bore. Tool hire companies normally have sets of alignment mandrels and one of them will fit your driven plate.

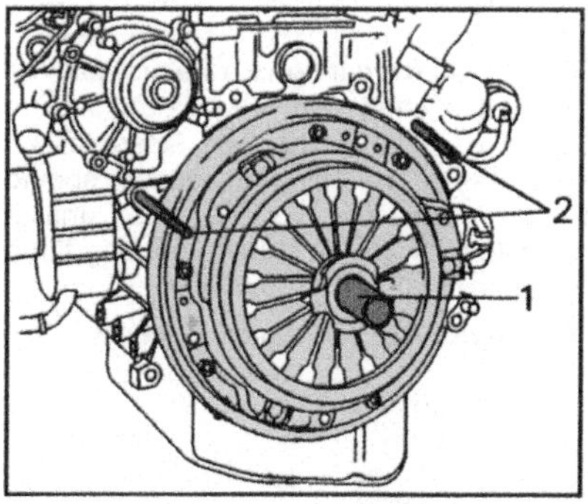

Fig. 2.4 – Fit the centring mandrel (1) into the driven plate when the bolts are tightened. Two studs are screwed into the cylinder block flange to guide the transmission during installation.

Offer up the clutch assembly to the flywheel and centre the disc by inserting the mandrel into the flywheel bearing and the splines. Tighten the bolts carefully, a little at a time, working in a diagonal manner, to a final torque setting of 2.0 kgm (14.5 ft.lb.).

Coat the clutch drive shaft and the outer face of the release bearing with a little Molycote BR2 grease (graphite grease) and refit the gearbox to the engine. Finally, adjust the clutch pedal clearance as described later on.

2.2. Servicing

The cover assembly – pressure plate and diaphragm spring – must not be dismantled. Replace, if necessary, with a complete assembly, but make sure to quote the year of manufacture and the chassis number.

Fig. 2.5 – Checking a clutch-driven plate for run-out between the centres of a lathe.

Inspect the driven plate and the linings, replacing the complete plate, if the lining material is worn down to the rivet heads. Again, specify the engine type and the model year, as different driven plates and clutch units are used. A driven plate with the linings contaminated with oil cannot be cleaned successfully and should also be replaced. All rivets should be tight and the torsion springs should be sound and unbroken. Also, check the condition of the driven plate splines. Clamp the driven plate between the centres of a lathe and apply a dial gauge to the outside of the plate, as shown in Fig. 2.5. Slowly rotate the disc and read off the dial gauge indication. Of more than 0.5 mm is indicated, fit a new disc.

Place a straight edge (steel ruler) over the friction face of the clutch unit and insert feeler gauges between the ruler and the surface. If the gap at the innermost spot on the friction face is no more than 0.3 mm, the plate can be re-used. Fig. 2.6 shows the check.

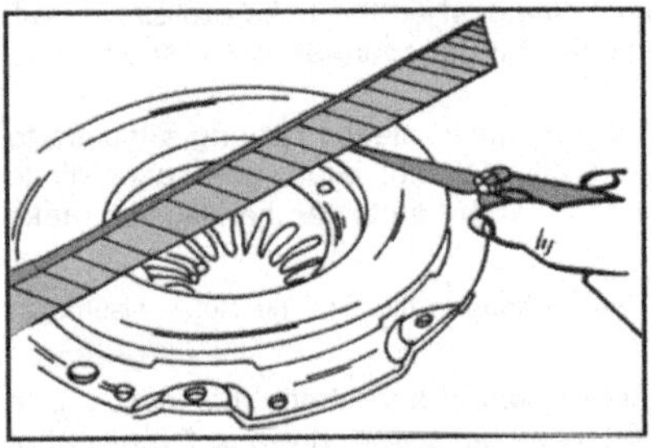

Fig. 2.6 – Checking the clutch pressure plate for distortion or wear at the inner edge.

Check the "fingers" of the diaphragm spring at the ends shown in Fig. 2.7 for wear. If more than 0.3 mm, replace the clutch. Also, make sure that all "fingers" are at the same level.

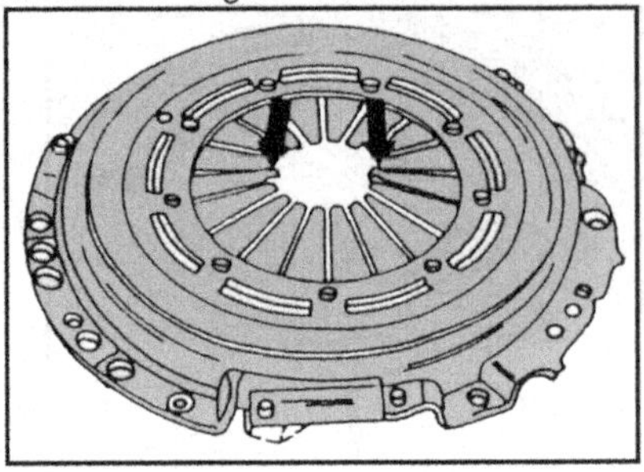

Fig. 2.7 – The "fingers" of the diaphragm spring are subject to wear. The ends must be at the same height.

2.2.1. CLUTCH RELEASE BEARING

The transmission must be removed to replace the clutch release bearing. Remove the release bearing from the inside of the gearbox bell housing and withdraw towards the front of the shaft. Refer to Fig. 2.3 when a release bearing of an MG5TU gearbox is dealt with. The bearing is attached to the clutch unit and we recommend that you can try to have it fitted in a workshop as special tools are necessary to remove the bearing, which is secured by a circlip. The release bearing is of the sealed type and must not be placed in any solvent.

The installation of a standard bearing is a reversal of the removal procedure. Coat the shaft and the bearing sleeve with Molycote BR2 grease. Grease the two sides of the release fork with MP grease. The release bearing should always be replaced at the same time as the clutch unit. It may, therefore, be possible to have the release bearing fitted where you buy the clutch unit.

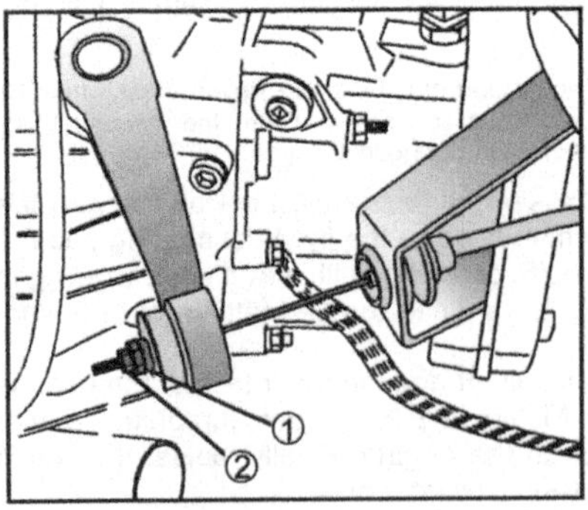

Fig. 2.8 – Details for the clutch pedal adjustment when the clutch release lever is "pushed".

1 Adjusting nut
2 Locknut

2.3. Clutch Cable Replacement

As already mentioned, the clutch release lever can either be "pulled" or "pushed". The removal and installation is, however, the same on both types. Figs. 2.8 and 2.9 show the attachment of the cable on the release lever on the two gearboxes. The other side of the cable is connected to the clutch pedal. Fig. 2.10 shows the complete mechanism. Proceed as follows.

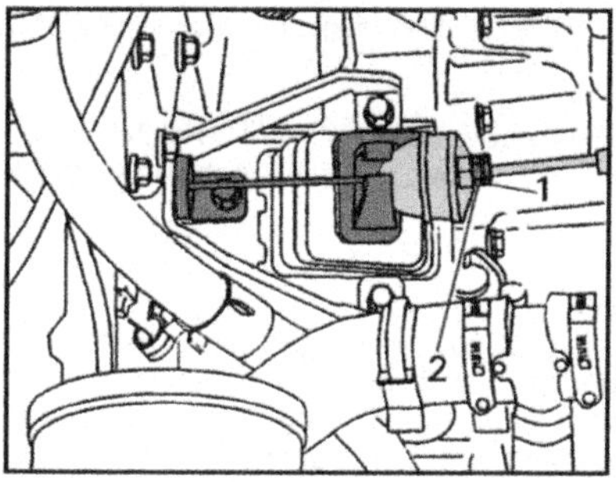

Fig. 2.9 – Details for the clutch pedal adjustment when the clutch release lever is "pulled".
1 Adjusting nut
2 Locknut

- Disconnect the battery and place the front end of the vehicle on the chassis stands. Remove the protective shield underneath the front of the vehicle.
- Unscrew the nut from the outside of the clutch release lever (Figs. 2.8 and 2.9). The nut is locked, i.e., 2 flat spanners are required.
- Pull the cable away from the gearbox until the outer sleeve is free. Depending on the type, disconnect the cable from the release lever and the bracket.
- Remove the lower trim on the driver's side to gain access to the clutch pedal. Pull the clutch pedal upwards until the cable can be disengaged. Direct the clutch cable end upwards and disengage it. Extract the clutch cable from the body panel from underneath the vehicle.

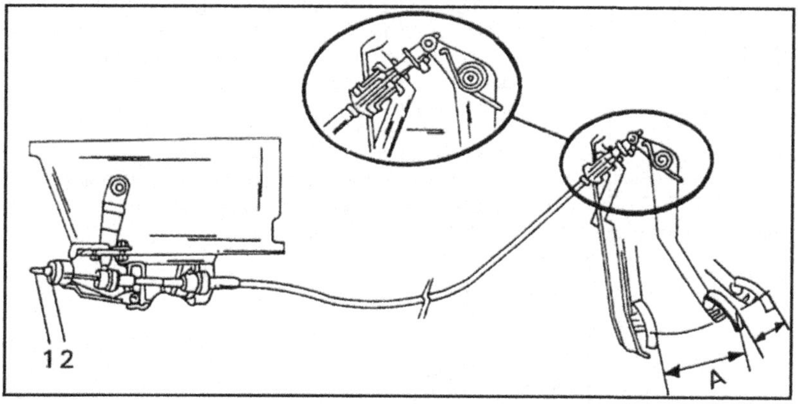

Fig. 2.10 – The routing of the clutch operating cable. The pedal travel is shown with "A".
1 Adjusting nut
2 Locknut

- Guide the clutch cable upwards until the clip can be removed from the pedal. The cable is secured by means of a bush, which in turn is secured by a clip. To remove the clip, use a screwdriver and push against in it the direction of the arrow in Fig. 2.11.

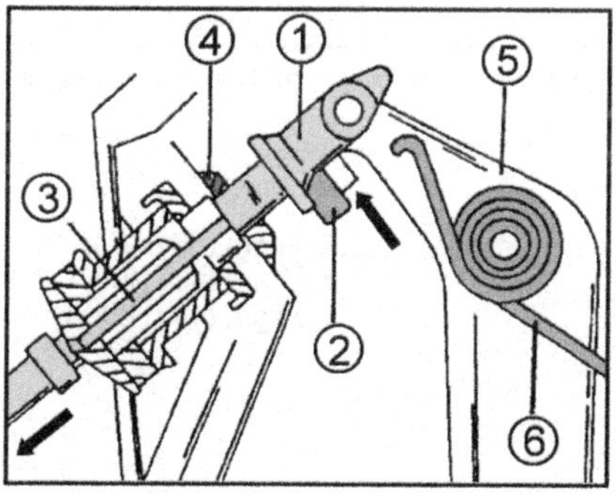

Fig. 2.11 – Attachment of the clutch cable at the upper end of the pedal. The clip must be pushed out in the direction of the arrow, using a screwdriver.
1 Bush
2 Clip
3 Clutch cable
4 Plastic ring
5 Clutch pedal
6 Return spring

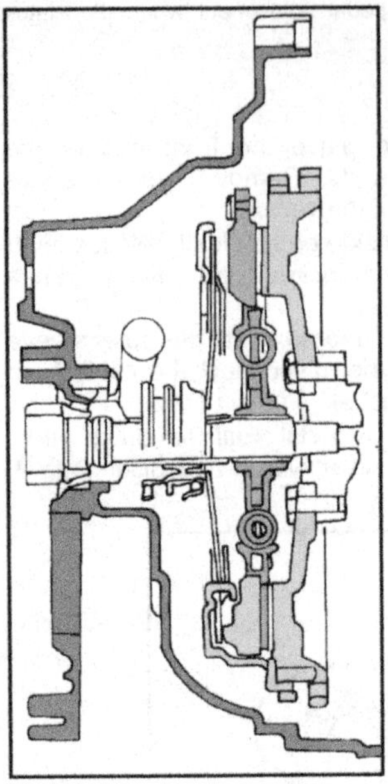

Fig. 2.12 – Sectional view of the clutch as fitted together with the ME5TU transmission.

The installation is a reversal of the removal procedure. The clutch pedal play must be adjusted as described below after the cable has been fitted.

2.4. Clutch Pedal Adjustment

First, ensure the type of clutch cable arrangement (gearbox) fitted to the vehicle, as the adjustment is not the same on all vehicles. Fig. 2.11 shows the clutch as fitted together with the HDi engine. As the clutch release bearing is in constant contact with the diaphragm spring, only the clutch pedal needs adjustment. Proceed as follows.

- Check that the clutch pedal is at the same height as the brake pedal.
- In the case of the "push" type, shown in Fig. 2.8, slacken the locknut (2) and rotate the adjusting nut (1) until both pedals have the same height. Tightening the nut will lift the pedal. Re-tighten the locknut and measure the pedal travel as described below.
- In the case of the "pull"-type arrangement, refer to Fig. 2.9, slacken the locknut (1) and rotate the adjusting nut (2) until both pedals have the same height. Tightening the nut will lift the pedal. Re-tighten the locknut and measure the pedal travel as described below.
- Push the pedal fully to the floor and apply a ruler as shown in Fig. 2.13. Release the pedal and measure dimension "X" as shown. This is also dimension "A" in Fig. 2.10. The pedal travel should be 180 mm, with a tolerance of +/−3 mm. It must never be less.

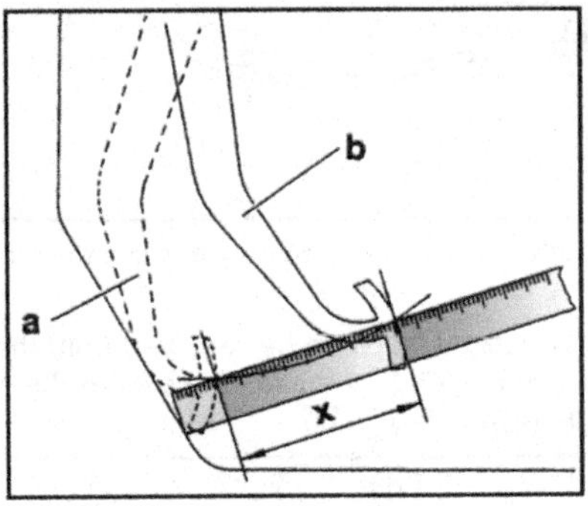

Fig. 2.13 – Measuring the pedal travel. Bring the pedal into position "a" and then into position "b". The difference is the clutch pedal travel, referred to the text.

2.5. Tightening Torques

Flywheel to crankshaft	5.0 kgm (36 ft.lb.)
Gearbox to engine	6.0 kgm (43 ft.lb.)
Clutch to flywheel	2.0 kgm (14.5 ft.lb.)

3. GEARBOX

3.0. Technical Data (All Models)

Type	5-speed transmission together with final drive in one housing. Fitted gearbox type depends on engine.

ME5TU and ML5TU Gearboxes

Identification	All transmissions have a number, commencing with "20KE" or "20LE" (ML5TU). The two end numbers are different and indicate the engine and carrying capacity.
Gearbox Ratios:	
- 1st gear	3.727:1
- 2nd gear	1.944:1
- 3rd gear	1.250:1 or 1.320:1
- 4th gear	0.880:1 or 0.969:1
- 5th gear	0.674:1, 1.077:1 or 0.814:1
- Reverse gear	2.000:1 or 3.154:1
Final drive ratio	5.61:1 or 4.470:1

Lubrication Oil

Filling capacity	1.85 litres
Oil change intervals	Every 40 000 miles
Recommended Oil	Total gearbox oil

MG5TU Gearbox

Identification	All transmissions have a number, commencing with "20KM". The two end numbers are different and indicate the engine and carrying capacity. Only fitted to models with 2.5- and 2.8-litre engines. The following gearboxes are fitted to 10Q, 14Q and 18Q vehicles, but note that some gearbox codes have changed
2.5-litre engine, DJ5-T9A, without turbo charger	Model 18Q (also with 2.8 litres)
2.5-litre engine, DJ5T-T8A, turbo charger	Model 14Q (also with 2.8 litres)
2.5-litre engine, DJ5T-T8A, turbo charger	Model 10Q (also with 2.8 litres)
2.5-litre engine, DJ5T-T8A, turbo charger	Model 18Q
2.5-litre engine, DJ5T-THZ, turbo charger	Model 18Q
2.5-litre engine, DJ5TED-THX, turbo charger	Different models

Gearbox Ratios:

- 1st gear	3.727:1
- 2nd gear	1.952:1
- 3rd gear	1.281:1
- 4th gear	0.953:1 or 0.884:1
- 5th gear	0.717:1
- Reverse gear	4.417:1
Final drive ratio	6.251:1, 5.231:1, 5.917:1 or 6.349:1

Lubrication Oil

Filling capacity:	
- With 1.9-/2.0-litre engine	1.85 litres
- With 2.5-/2.8-litre engine	2.75 litres
Oil change intervals	Every 40 000 miles
Recommended Oil	Total gearbox oil

Note: A type identification plate is fitted to the transmission, but we would like to point out the differences Again, all transmissions are considered.

- A type identification plate is fitted to the side of the transmission as shown on the L.H. side of Fig. 3.1.
- In the R.H. view of Fig. 3.1, a type ML5TU transmission is shown. The type identification plate is attached at position (1). Additionally, the serial number and the transmission number are stamped into the transmission at position (2).

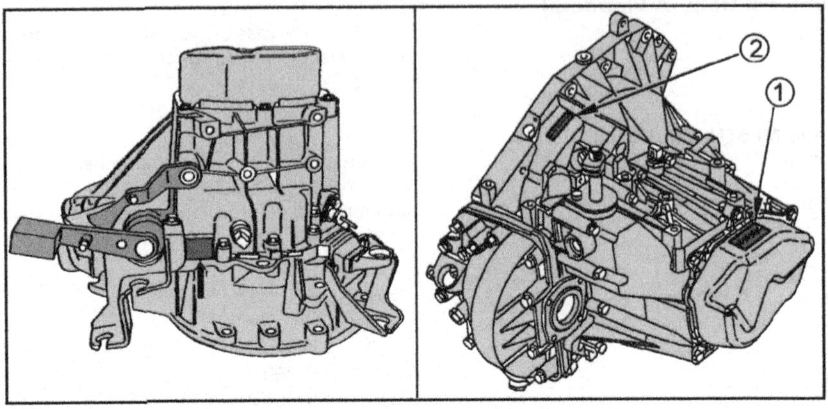

Fig. 3.1 – Location of the type identification plate(s) in the case of the MG5TU transmission (L.H. side) and the ML5TU transmission (R.H. side).

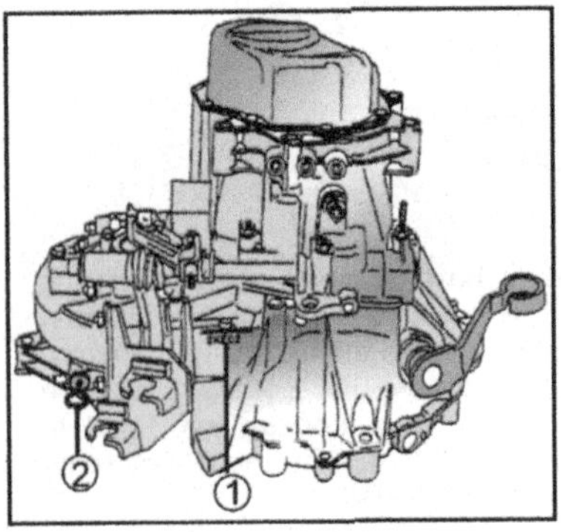

Fig. 3.2 – The identification plate (1) in the case of an ME5TU transmission is located on the same side as the oil dipstick (2).

- Fig. 3.2 shows the location of the identification details in the case of an ME5TU transmission. The transmission number is stamped into the transmission at position (1). To find the number, use the oil dipstick for orientation.

3.1. Gearbox – Removal and Installation

Although there are some differences during the removal and installation of the various gearboxes, all models are covered below. The transmission is removed without the engine. Note the following before commencing removal.

- There are two oil drain plugs fitted to the ME5TU gearbox, as can be seen on the L.H. side of Fig. 3.3. One drain plug (1) is fitted to the actual transmission housing and one plug (2) in the final drive housing (differential housing).
- The MG5TE gearbox has one oil drain plug, situated at the bottom of the transmission. This version is shown in the R.H. view of Fig. 3.3.
- In the case of the ML5TU transmission, two plugs are present, as can be seen in Fig. 3.4. The oil drain plug (1) is fitted to the bottom of the transmission. Plug (2) must be removed to check the oil level.

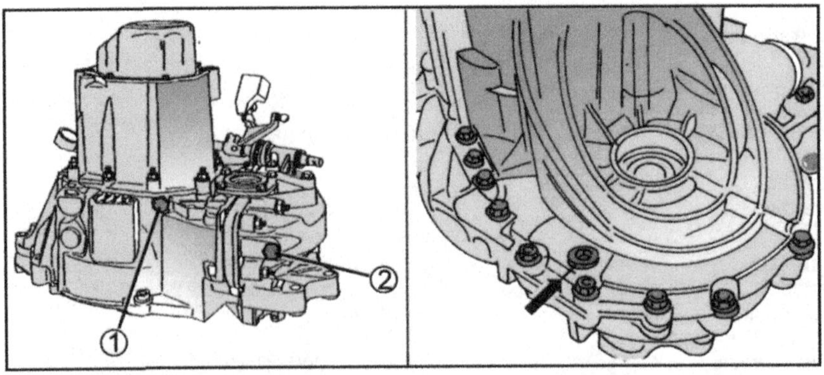

Fig. 3.3 – The location of the drain plugs on the various transmissions. Refer to the text.

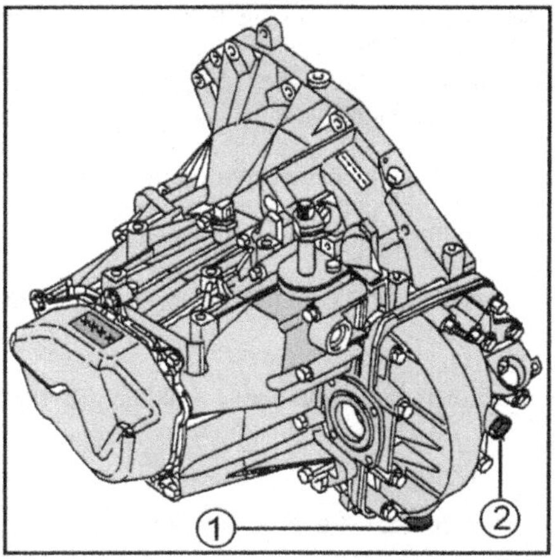

Fig. 3.4 – The transmission drain plug (1) and the oil level check plug (2) on the ML5TU transmission. Plug (2) must be removed to refill or top-up the transmission.

Remove the transmission as follows, noting that again all different types are covered.

- Disconnect the negative battery cable.
- Remove the protective panel underneath the front of the vehicle.

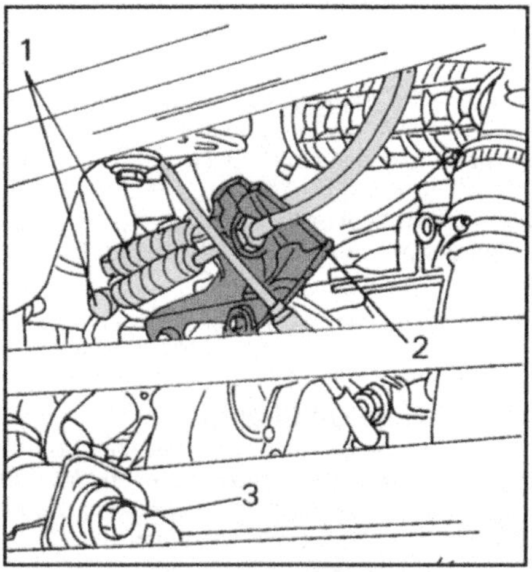

Fig. 3.5 – The attachment of the gearchange cables on the transmission.
1 Gearchange cables
2 Cable abutment bracket
3 Torque strut

- Drain the gearbox oil (see above and description later on). Slacken the wheel bolts and the drive shaft nuts as explained during the removal of the drive shafts, place the front end of the vehicle on chassis stands and remove the wheels. The two drive shafts are now removed as described in Section 4.1.
- Remove the air cleaner together with the air intake hoses. In the case of a turbo-charged engine, remove the hose/pipe connections between the turbo charger and the intercooler (charge air cooler, if fitted).
- Disconnect the clutch cable as already described.
- Unscrew the battery earth cable.
- Separate the front exhaust pipe from the exhaust manifold flange. Use a piece of wire and tie the pipe to the bottom of the vehicle. Disconnect the speedometer drive cable and pull it out of the gearbox.

The following operations are carried out by referring to Fig. 3.5. Remove the two gearchange cables (1), unscrew the bracket (2) and separate the torque strut (3).

- Remove the protective panel underneath the flywheel housing. Disconnect the cables from starter motor and remove the starter motor.
- Disconnect the cables from the reversing light switch.
- Attach the engine to a suitable hand crane or hoist, using ropes or chains and lift the assembly until you can feel that it is under tension. It is also possible to place a hydraulic jack underneath the engine (wooden plank between jack and oil sump) to lift the power unit. Place a mobile jack under the transmission to support the transmission from below in preparation for the next operation.

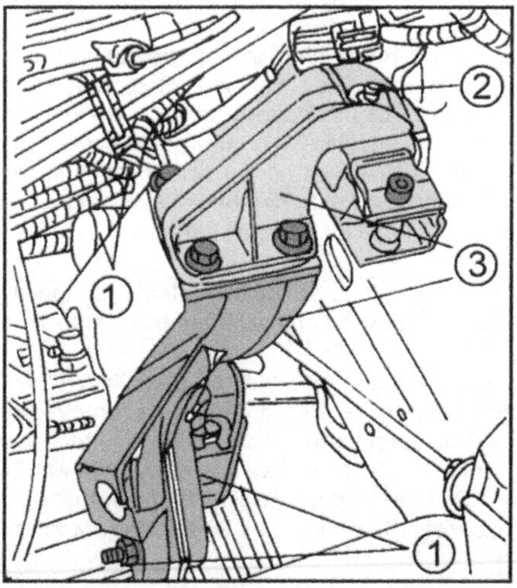

Fig. 3.6 – The gearbox mounting.
1 Bolts, 5.0 kgm
2 Nut, 9.0 kgm
3 Mounting

- Remove the gearbox mounting by referring to Fig. 3.6. First, remove the four bolts (1) and then the nut (2). Check that the engine is supported by the jack or the hoist and remove the two parts of the gearbox mounting.
- Remove the bolts between engine and gearbox at the top and bottom, after checking that the gearbox is well-supported on the jack.
- Withdraw the gearbox from the engine. A helper is required to keep the gearbox on the jack and to guide it out.

The installation is a reversal of the removal procedure, noting the following points.

- Coat the clutch shaft splines and the guide sleeve for the clutch release bearing with a little graphited grease. Check that the clutch release fork is resting against the release bearing.
- To facilitate the installation of the gearbox, prepare two studs with suitable thread. Cut a slot for a screwdriver into one end and screw the studs at the top on the right-hand side and at the bottom on the left-hand side into the engine mounting flange.
- Lift the engine in position on the jack and guide the clutch shaft into the clutch-driven plate. Push the gearbox against the engine and over the two studs. Fit some of the bolts and tighten them until the gap between gearbox and engine is nearly closed. Finally, unscrew the two studs and replace with the remaining bolts. Tighten all bolts evenly to 6.0 kgm (43 ft.lb.).
- Fit the gearbox mounting with reference to Fig. 3.5. Tighten bolts (1) to 5.0 kgm (36.5 ft.lb.) and the nut (2) to 8.0 kgm (58 ft.lb.).
- Tighten the torque strut (3) in Fig. 3.5 with 9.0 kgm (65 ft.lb.).
- Refit the drive shafts as described in Section 4.1., connect the clutch cable and finally refill the gearbox with oil.

Models With Four-Wheel Drive
The removal is carried out in a similar manner, with the difference that the propeller shaft must be disconnected at the rear of the gearbox.

3.2. Gearbox Repairs

Many special tools are required to dismantle, assemble and adjust the transmission. Additionally, there are, of course, two transmissions to be considered. The only operation we recommend is the replacement of the oil seals in each side of the transmission, which is described in the section dealing with the drive shafts.

A damaged or faulty gearbox should be taken to a workshop for overhaul or try to obtain an exchange gearbox. In this case, you will need the usual chassis number, model year, model, etc., to ensure that the correct gearbox is fitted.

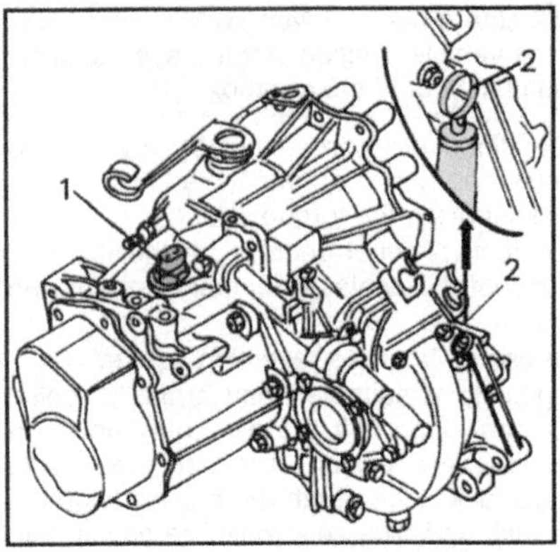

Fig. 3.7 – On the ME5TU gearbox, the reversing light switch (1) is used to fill the gearbox. The oil dipstick (2) is used for the oil level check.

3.3. Gearbox Oil Level and Oil Change

Only the interval to change the gearbox oil is the same on all gearbox types, i.e., the oil should be changed every 40 000 miles. When changing the oil and also when checking the oil level, proceed as follows for different gearbox types.

ME5TU Gearbox
This gearbox has an oil drain plug in the actual gearbox housing and one in the final drive housing. Both plugs are located next to each other at the bottom of the transmission housing, just below the drive shaft exit on the cover side of the gearbox. To check the oil level, an oil dipstick is fitted at the position (2) in Fig. 3.7 into the final drive housing. Checking is simple. Withdraw the dipstick and read off the oil level. Correct the oil level as described below.

Change the gearbox oil as follows.

- Unscrew the two drain plugs and allow the gearbox oil to drain into a suitable container (capacity approx. 5 pints). Warm gearbox oil will drain easier. Clean and refit the plugs after the gearbox is empty.
- Fill the gearbox with the recommended oil (SAE 75W/80W). The reversing light switch (1) in Fig. 3.7 must be removed to fill the gearbox (first disconnect the cable). Use a funnel and fill the gearbox until the level remains steady. An empty gearbox will take 1.85 litres. After a while, re-check the level with the dipstick and refit the switch if all is well. Do not forget the cable.

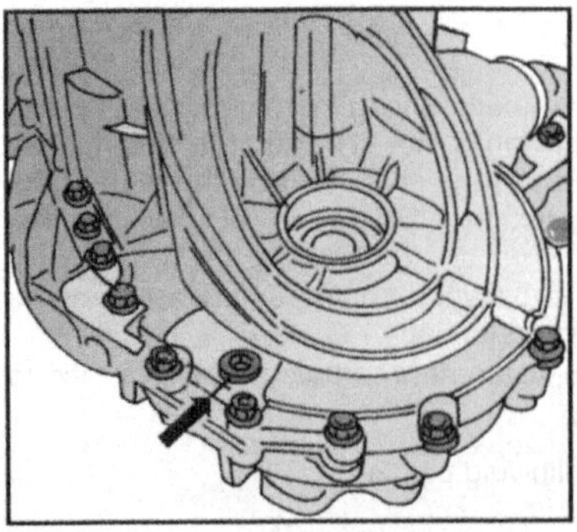

Fig. 3.8 – The arrow shows where the oil drain plug of an MG5TU gearbox is located.

MG5TU Gearbox
Only one drain plug is fitted to the bottom of the gearbox, at the position shown in Fig. 3.8. An oil level checking/filler plug is fitted,

but the gearbox has been modified since its introduction. Originally fitted gearboxes (not applicable with a 2.0-litre engine) have an oil dipstick at the position shown in Fig. 3.9, secured by a screw. To check the oil level, unscrew the dipstick and pull it out. Read off the oil level and correct it if necessary. Later, a plug with a square opening is fitted below the drive shaft exit in the gearbox. To check the oil level, unscrew the plug with a square wrench and check that the oil is at the lower edge of the plug opening. Correct the oil level. An oilcan, filled with oil, is recommended.

MG5TU Gearbox

Only one oil drain plug is fitted, as shown in Fig. 3.4. A further plug is used to check the oil level and to top-up the oil. To check the oil level, remove the plug (2) and check that the oil is up to the lower edge of the plug bore. Use a grease gun filled with transmission oil (clean it first) to top-up or refill the transmission.

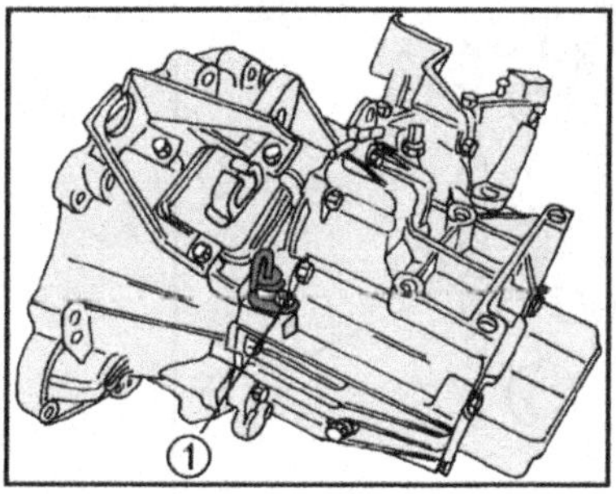

Fig. 3.9 – Oil dipstick and filler opening (1) of an MG5TU gearbox on early gearboxes. Later gearboxes have a plug in the side of the transmission.

The oil change is carried out as described for the other gearboxes.

3.4. The Gearchange

The gearchange is actuated by means of cables, one selector cable and one gearchange cable. The cables can be recognised on the ball joint connections; the gear change cable has a larger ball joint. Only the selector cable can be adjusted. Fig. 3.10 shows the arrangement of the gearchange elements.

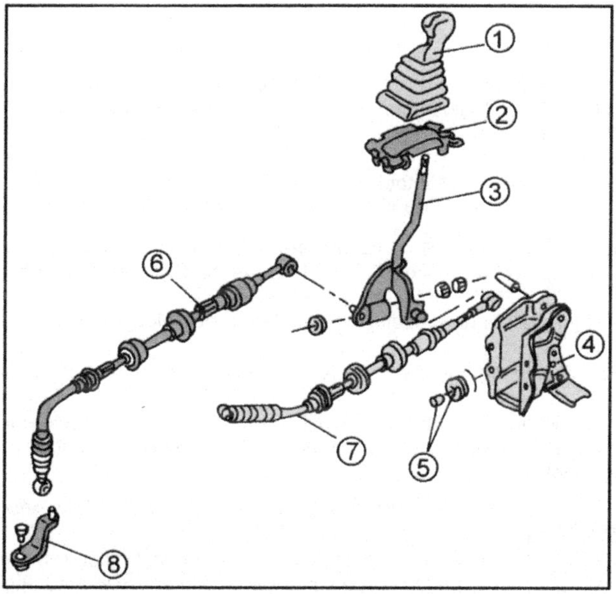

Fig. 3.10 – The component parts of the gearchange mechanism.

1 Gearchange lever
2 Gearchange leverbearing
3 Gear change lever
4 Cable bracket ("2" in Fig. 3.5)
5 Elastic mounting
6 Gear change cable
7 Gear selector cable
8 Lever on gearbox

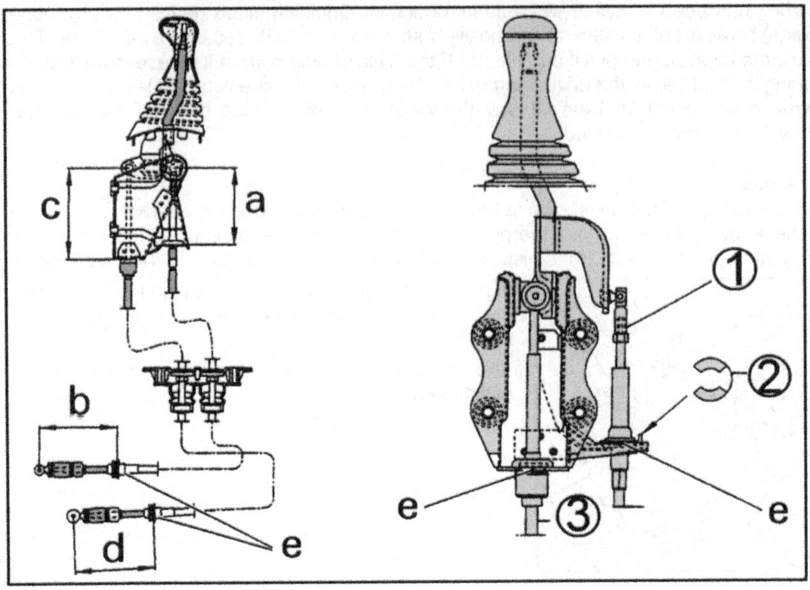

Fig. 3.11 – Cable gearchange together with the adjusting dimensions. Dimensions "a"–"e" are not the same on all transmissions (see below).
1 Gear selector cable (adjustable)
2 Securing clip
3 Gear change cable (not adjustable)

MG5TU transmission:
a = 123 +/– 5 mm (ajustable)
b = 123 +/– 1.5 mm
c = 144 +/– 1.5 mm
d = 129 +/- 1.5 mm (adjustable)
e = Cable attachment (2 clips)

ME5TU transmission:
a = 123 +/– 5 mm (adjustable)
b = 123 +/– 1.5 mm
c = 144 +/– 1.5 mm
d = 139 +/– 1.5 mm (adjustable)
e = Cable attachment (2 clips)

ML5TU transmission:
a = 123 +/– 5 mm (adjustable)
b = 147 +/– 1.5 mm
c = 144 +/– 1.5 mm
d = 123 +/– 1.5 mm (adjustable)
e = Cable attachment (2 clips)

The removal and installation of the cables is fairly easy. The cables are secured on the side of the gearbox and the gear lever with "C" -shaped clips on a bracket. Fig. 3.11 shows the two cables together with the fitted dimensions.

After a cable has been replaced, adjust the gearchange mechanism as described below. A steel ruler is required for the adjustment.

- Place the gear change lever into the neutral position and check if the dimension between the arrows in Fig. 3.12 is exactly 304 mm.
- If necessary, disconnect the ball joint connection (1) in Fig. 3.11, slacken the locknut at the cable end and turn the end piece until the dimension is obtained. Several attempts will be necessary to achieve success. Tighten the locknut after adjustment.

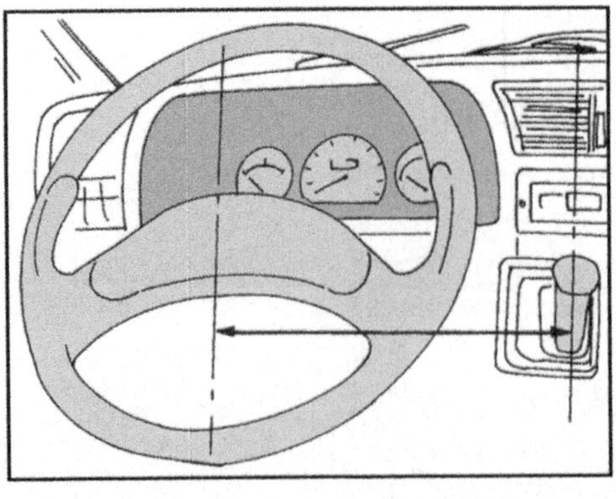

Fig. 3.12 – The dimension is measured between the arrows. The correct dimension is 304 mm.

If one or the other cable is replaced, adjust it to the correct dimension before installation. Refer to Fig. 3.10 and follow the given dimensions. Measure between the arrows.

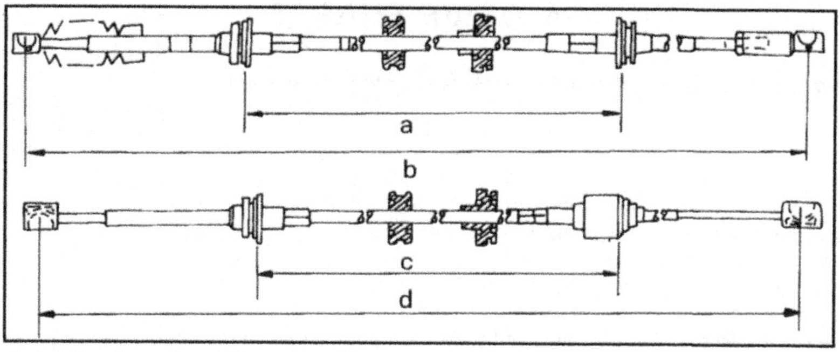

Fig. 3.13 – The installation dimensions of the cables. The upper view shows the selector cable, and the lower view shows the gear change cable. The dimensions are given for the MG5TU transmission. The other transmissions are listed below.

MG5TU transmission:

a = 750 mm +/– 2 mm	c = 690 mm +/– 2 mm
b = 996 mm +/– 7 mm	d = 963 mm +/– 3 mm

ME5TU transmission:

a = 750 +/– 2 mm	c = 752 +/– 2 mm
b = 1006 +/– 7 mm	d = 1035 +/– 3 mm

ML5TU transmission:

a = 745 +/– 2 mm	c = 720 +/– 2 mm
b = 991 +/– 7.75 mm	d = 1011 +/– 3 mm

3.5. Gearbox – Tightening Torque Values

Drive shaft nuts:	
- Models 10Q and 14Q	45.0 kgm (324 ft.lb.)
- Model 18Q	50.0 kgm (360 ft.lb.)
Wheel Bolts:	
- Models 10Q and 14Q	16.0 kgm (115.5 ft.lb.)
- Model 18Q	18.0 kgm (130 ft.lb.)
Nut of Lower Suspension Arm Ball Joint:	
- Models 10Q and 14Q	13.0 kgm (94 ft.lb.)
- Model 18Q	24.0 kgm (173 ft.lb.)
Track rod ball joint to steering lever	7.0 kgm (50.5 ft.lb.)
Intermediate bearing locking nuts	1.0 kgm (7.2 ft.lb.)
Brake caliper to steering knuckle	21.0 kgm (151 ft.lb.)
Brake caliper guide bolts	3.5 kgm (25 ft.lb.)

4. DRIVE SHAFTS

Note: Observe different tightening torques applicable to the wheel bolts.

The L.H. and R.H. drive shafts are of different lengths. Drive shafts always look the same. When a new shaft is purchased, make sure you ask for the correct shaft. The following notes will help you with identification. "Loctite" thread locking compound will be required for the installation of the shaft.

The fitted gearbox is also responsible for the type of drive shaft and the outer diameter of the CV joint on the transmission side (86 mm or 92 mm in the case of an MG5TE gearbox) and on the wheel side (100 mm or 113 mm in the case of the MG5TE gearbox). The shafts are identified with a code in order to allocate the correct shaft to the vehicle. We, therefore, recommend to take the old shaft to the spare part counter to obtain the correct shaft.

Finally, we must add that the CV joints of the two shafts are different. A ball-type CV joint is fitted to the wheel side. A tri-axe Rzeppa joint is fitted on the gearbox side of all models.

An intermediate bearing is used to locate the R.H. drive shaft. The nut at the end of the shaft is tightened with a very high torque (again not the same on all models) and is subsequently secured in position by peening the shoulder of the nut into a groove in the shaft.

Fig. 4.1 shows the two drive shafts.

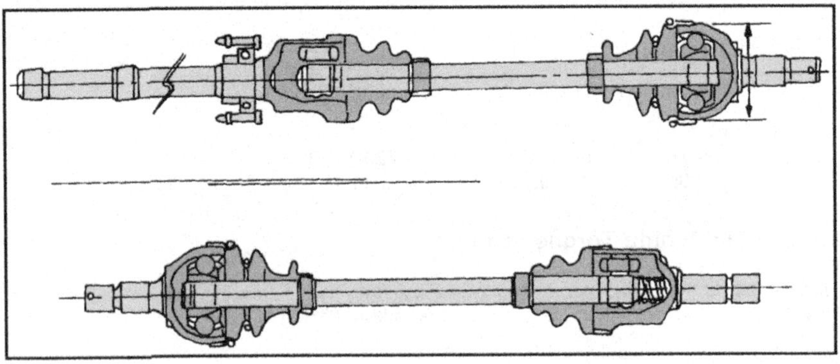

Fig. 4.1 – View of the two drive shafts. The upper view shows the R.H. shaft with support bearing, and the lower view shows the L.H. drive shaft. The support bearing is not the same for all transmissions.

4.0. Technical Data

Type	See above
Intermediate bearing dimensions	35 × 62 × 16 mm
Diameter of CV joint, gearbox side:	
- With ME5TU and ML5TU gearboxes:	86 mm
- With MG5TE gearbox	92 mm
Diameter of CV joint, wheel side:	
- With ME5TE and ML5TU gearboxes:	100 mm
- With MG5TE gearbox	100 or 113 mm
Diameter of Axle Stump:	
- With ME5TE and ML5TU:	30 mm
- With MG5TE gearbox:	30 mm (Q10/Q14) or 35 mm (Q18)
Lubrication grease	Total Multi-Grease (G6)

4.1. Removal and Installation

It should be noted that the oil seal for the drive shaft in the gearbox housing must be replaced, if a shaft is removed, irrespective of what type of repairs are carried out. A suitable mandrel is, therefore, required to fit the new seal. Also note the following.

- The nut for the lower suspension ball joint has a different torque setting within the vehicle range.
- A 50-mm socket is required to slacken and tighten the axle shaft nut, but again, a different size may be required, depending on the model.
- The tightening torque of the drive shaft nut is very high. *Never try to slacken or tighten the nut when the vehicle is on chassis stands.*

Remove a shaft as follows.

- Disconnect the battery earth cable.
- Apply the handbrake and engage the first gear.
- Slacken the wheel bolts. Do not lift the vehicle at this stage.
- Remove the hub grease cap, using a small chisel or a strong screwdriver. If difficulties are experienced, remove the wheel to remove the cap, but refit the wheel and lower the vehicle before the next operation.
- Slacken the drive shaft nut. The vehicle can now be placed on chassis stands and the nut removed from the shaft end. A suitable lever as shown in Fig. 4.1 can be used to slacken the nut if the wheel has been removed.

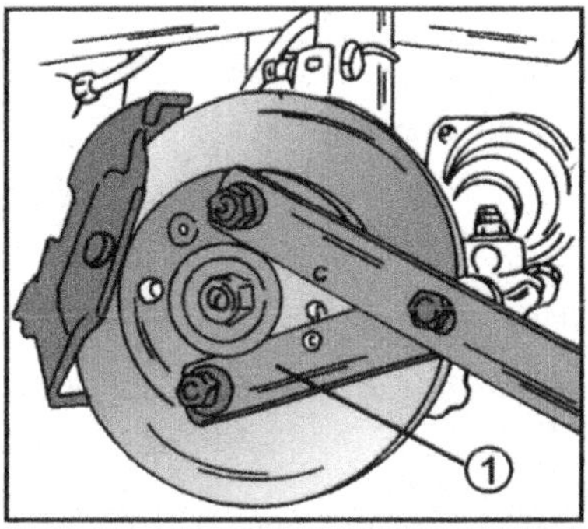

Fig. 4.2 – Counterholding a front wheel hub when slackening or tightening the shaft nut. A suitable lever (1) can be made up.

- Partially drain the gearbox oil. Depending on the transmission, one or two drain plugs must be removed. Allow about 1 litre of oil to drain in a suitable container. This will prevent oil from running out of the gearbox opening. If required, drain the gearbox completely.
- Remove the track rod ball joint nut and separate the joint from the lever on the steering knuckle (refer to the "Steering" chapter).
- Remove the nut securing the suspension ball joint and separate the joint in the manner shown in Fig. 4.3.

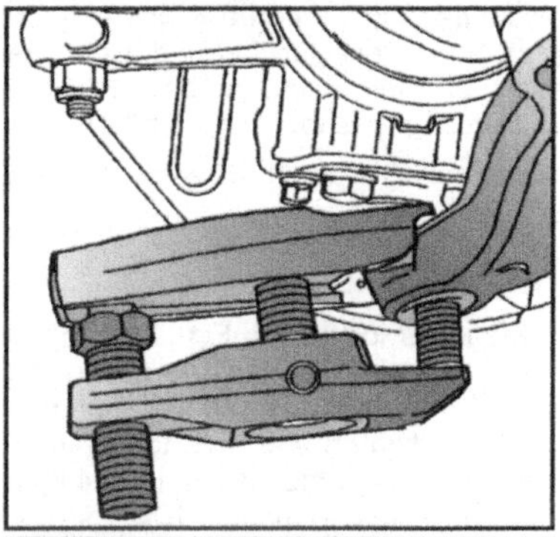

Fig. 4.3 – Separating the suspension ball joint at the bottom of the steering knuckle with a ball joint. The nut can remain on the stud to protect the thread.

The remaining removal depends on the side in question.

L.H. Drive Shaft

- Pull the steering knuckle towards the outside until the shaft is free of the wheel hub. The shaft can now be pulled out of the gearbox.

Fig. 4.4 – The arrows show where the support bearing for the R.H. drive shaft is attached. After removal of the nuts, rotate the bolt heads as described.

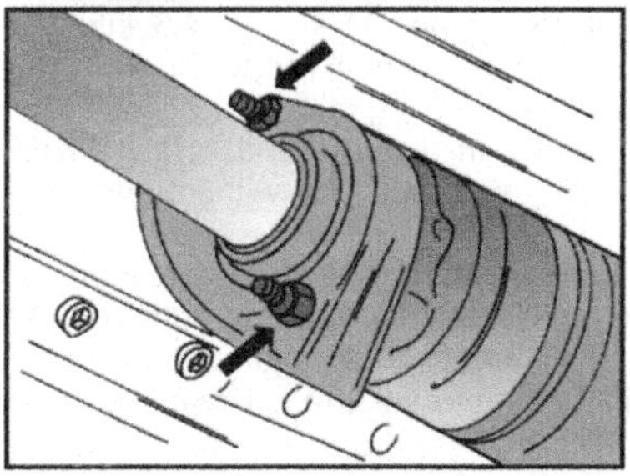

R.H. Drive Shaft

- Pull the steering knuckle towards the outside until the shaft is free of the wheel hub.

Remove the two nuts in Fig. 4.4 and turn the bolt heads by a quarter of a turn from its original position. The shaft can now be removed.

Fig. 4.4 – The arrows show where the support bearing for the R.H. drive shaft is attached. After removal of the nuts, rotate the bolt heads as described.

Both Shafts

- Pull the wheel at the bottom towards the outside until the shaft is free from the inside of the wheel hub. When the shaft is free, remove it from the gearbox opening.
- The oil seal in the gearbox opening should be replaced. Oil leaks are possible if this precaution is ignored.

The installation of the drive shaft is carried out as follows.

- Clean the surrounding of the gearbox opening and drive a new oil seal into position until the outside is flush. The oil seal lip should be coated with gearbox oil. Take care not to damage the seal.
- Insert the installation sleeve, normally supplied with the new seal, into the differential opening as shown in Fig. 4.5 and slide the inner end of the shaft into the gearbox, engaging the splines on shaft and differential side gear. After the shaft is in position, grip the "ears" of the sleeve and withdraw it.

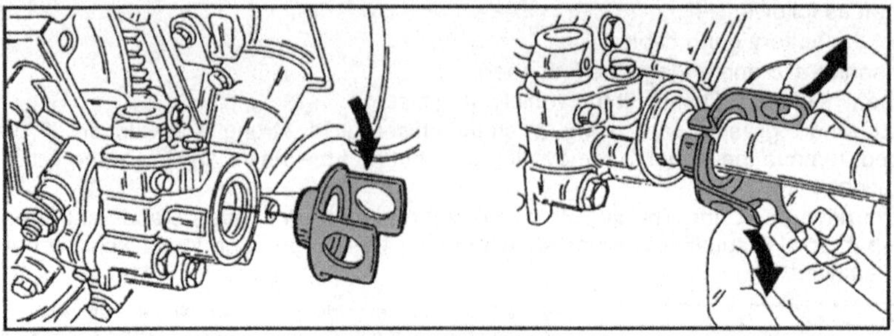

Fig. 4.5 – Fitting the installation sleeve in the L.H. view and removal of the sleeve after installation of the shaft in the R.H. view.

- If both shafts have been removed, carry out the same operations on the other side.

The remaining installation is different for the two shafts.

L.H. Shaft

- Pull the knuckle towards the outside as far as it will go and insert the shaft end into the steering knuckle.
- Push the steering knuckle towards the inside until the shaft end protrudes from the wheel hub. Fit the suspension ball joint and the track rod ball joint as described below.

R.H. Shaft

- Lubricate the outer bearing race and the oil seal of the intermediate bearing with grease and engage the drive shaft with the intermediate bearing. Bring the bearing in the correct position and insert the drive shaft into the differential side gear and the wheel hub. When in position, rotate the bolt heads by a quarter of a turn against the outer bearing race. Tighten the two securing nuts to 1 kgm (7 ft.lb.).

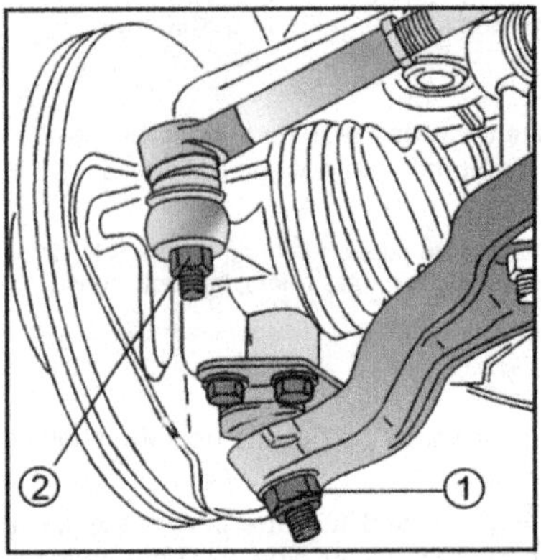

Fig. 4.6 – Always replace the two nuts (1) and (2). Note the tightening torque of nut (2) is not the same on all models.

Both Shafts

- Fit the ball joint stud of the lower joint (2) in Fig. 4.6 into the steering knuckle and fit a *new* nut. Tighten the nut to the correct torque, depending on the fitted shaft. If the shaft end has a diameter of 30 mm, tighten the nut to 13 kgm (94 ft.lb.); if the diameter is 35 mm, tighten the nut to 24 kgm (173 ft.lb.). The latter shafts are fitted to models 18Q.
- Insert the track rod ball joint stud (1) in Fig. 4.6 into the steering lever and fit and tighten a *new* nut to 7.0 kgm (50 ft.lb.).
- Coat the drive shaft thread with "Loctite" and screw a ***new*** nut onto the shaft as far as it will go, without the final tightening.
- Fit the wheel and lower the vehicle onto the ground. Apply the handbrake. Tighten the drive shaft nut with a 50-mm socket. Again, the drive shaft diameter is used to apply the torque. Either tighten to 45 kgm 325 ft.lb.), if the diameter is 30 mm, or 50 kgm (360 ft.lb.), if the diameter is 35 mm.
- After the nut is tightened, use a blunt chisel and peen the nut shoulder into the shaft. The chisel must be applied vertically from above to avoid slackening of the nut.
- Fit a new grease cap.
- Tighten the wheel bolts (note the torque setting).
- Fill the gearbox with oil or correct the oil level, if the oil has only been drained partially.

4.2. Drive Shaft Repairs

Drive shaft repairs are always difficult, if the CV joint is worn or if the rubber gaiter is torn, we recommend taking the drive shaft to a workshop and having it overhauled or fit a new shaft. If dirt has entered through a torn rubber gaiter, it will have entered the CV joint, requiring the dismantling the greasing of the latter. The replacement of the rubber gaiter on the wheel side also requires the replacement of the gaiter on the transmission side.

Check the condition of the rubber gaiters at regular intervals to prevent road dirt from entering the joints.

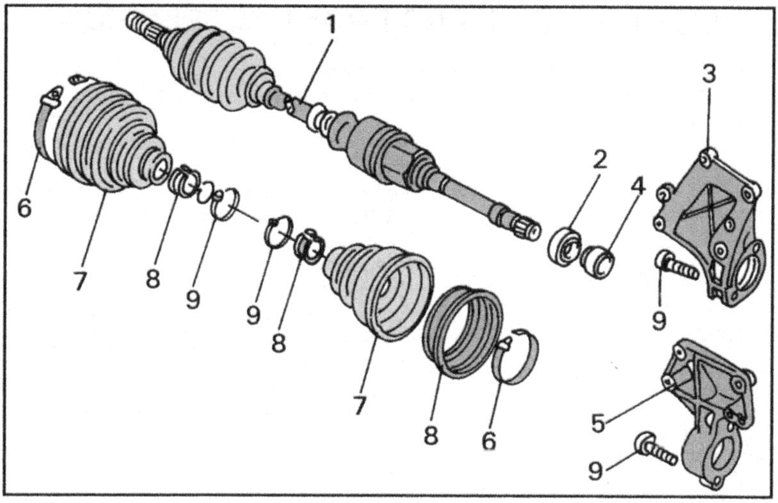

Fig. 4.7 – Only the illustrated parts of the drive shaft should be replaced.

1 R.H. drive shaft
2 Ball bearing
3 Support bracket, MG5TU
4 Bearing spacer
5 Support bearing bracket (ME5TU/ML5TU)
6 Large gaiter clip
7 Rubber gaiter
8 Sleeve for gaiter
9 Bolt, 1.0 kgm

4.3. Replacing the Support Bearing

The bearing can be replaced after the removal of the R.H. shaft. Remove the bearing bracket from the cylinder block. If the bearing has not come off together with the drive shaft, remove it together with the oil seal from the bearing bracket. Fit the new bearing and the oil seal, refit the bearing bracket to the cylinder block and refit the drive shaft. The bearing must be greased with MP grease.

A different bearing bracket is used, depending on which type of transmission is fitted. Fig. 4.7 shows an exploded view of the shaft for reference.

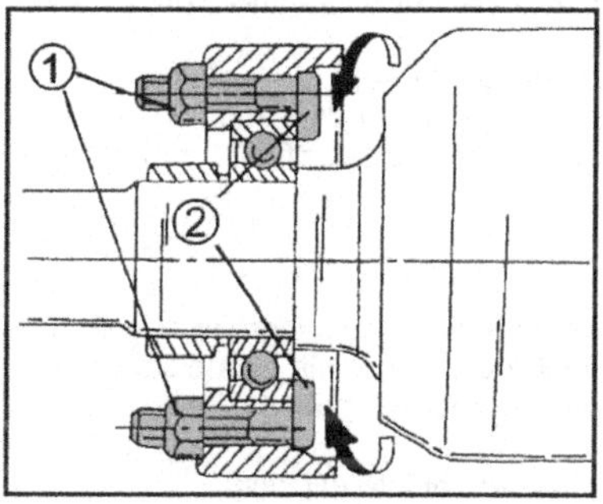

Fig. 4.8 – Sectional view of the support bearing attachment (R.H. shaft).
1 Nuts, 1.0 kgm
2 Bolt heads

The nuts securing the outer bearing race are tightened to 1.0 kgm (7.2 ft.lb.). Fig. 4.8 shows a sectional view of the support bearing. The illustration shows the bolt heads in the locked position after they have been rotated by a quarter of a turn.

4.4. Drive Shafts – Tightening Torque Values

Drive shaft nuts:	
- Models 10Q and 14Q	45.0 kgm (324 ft.lb.)
- Model 18Q	50.0 kgm (360 ft.lb.)
Wheel Bolts:	
- Models 10Q and 14Q	16.0 kgm (115.5 ft.lb.)
- Model 18Q	18.0 kgm (130 ft.lb.)
Nut of Lower Suspension Arm Ball Joint:	
- Models 10Q and 14Q	13.0 kgm (94 ft.lb.)
- Model 18Q	24.0 kgm (173 ft.lb.)
Track rod ball joint to steering lever	7.0 kgm (50.5 ft.lb.)
Intermediate bearing locking nuts	1.0 kgm (7.2 ft.lb.)
Brake caliper to steering knuckle	21.0 kgm (151 ft.lb.)
Brake caliper guide bolts	3.5 kgm (25 ft.lb.)

5. FRONT AXLE AND FRONT SUSPENSION

5.0. Technical Data

Type	Independent suspension with McPherson spring struts, lower suspension arms, integral shock absorbers and coil springs. With stabiliser bar, depending on model. Stabiliser bar mounting: mounting clamps and rubber bushes, bolted to crossmember and connected to suspension arms with links.
Front springs	Colour coded. Only fit springs with original colour code. The 1st marking denotes vehicle model, and 2nd marking refers to weight category. Vehicle model and model year must be quoted when fitting new springs.
Stabiliser bar (if fitted)	24 mm (0.95 in.) diameter. Fitted to models 18Q and 10Q/14Q with turbo diesel engine.
Shock absorber identification:	
Models 10Q/14Q	Black marking
Model 18Q	Blue marking
Wheel Alignment	
Caster, not adjustable (laden):	1° +/– 30'
Caster, not adjustable (empty):	0° +/– 45'
Camber, not adjustable	0° +/– 30'
Toe-in, adjustable	0 +/– 1 mm (0 +/– 0.04 in.)
Checking conditions	Tyre pressures correct, vehicle empty or Laden (see above)
Front Wheel Bearings	
Type	Wheel hub with two taper roller bearings, pressed into steering knuckle
End float of bearings	0.02–0.10 mm (0.008–0.004 in.)
Means of adjustment	Spacer sleeves between 21.29 and 21.97 mm, available in various sizes, in steps of 0.04 mm between sizes.

5.1. Regular Maintenance Operations

The suspension ball joints between the suspension arms and the wheel bearing housings, also known as steering knuckles, require no maintenance. Rubber dust caps prevent the entry of dirt or moisture. Entry of dust has a grinding effect; entry of moisture leads eventually to rust formation. During regular maintenance work, you can check the ball joints as follows.

- Place the front of the vehicle on chassis stands and turn the steering wheel into full lock.
- Check the rubber dust caps on both sides for cuts or other damage. A damaged dust cap must be replaced by fitting a new ball joint. Apart from the fact that the dust caps cannot be replaced individually, there is the possibility that dirt has already entered the joint, thereby leading to wear.

Shock Absorbers and Spring Struts

In Section 5.2.1., you will find certain instructions on how to check shock absorbers. Follow the advice.

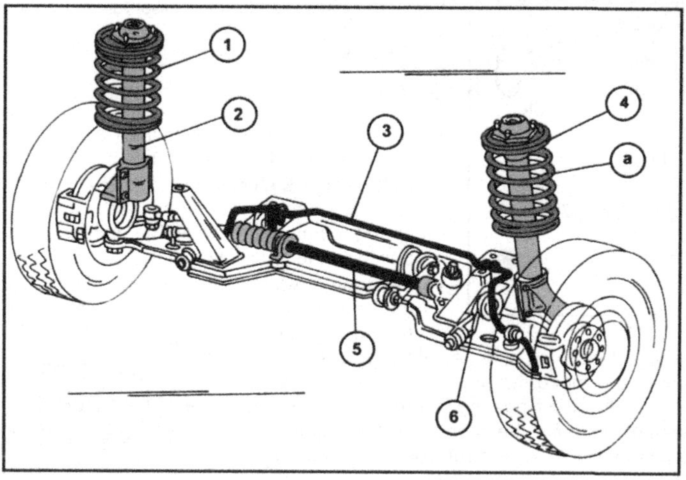

Fig. 5.1 – View of the assembled front suspension. The coil springs are marked with a spot of paint at position "a".

1 Coil spring
2 Spring strut
3 Stabiliser bar
4 Rebound rubber
5 Suspension arm
6 Steering unit

5.2. Front Spring Struts

5.2.0. REMOVAL

The following text describes the removal of the complete spring strut without the suspension arms and the drive shaft. The coil spring must be compressed before a spring strut can be dismantled. The wheel can remain on the hub or is removed to improve access to the various mounting points.

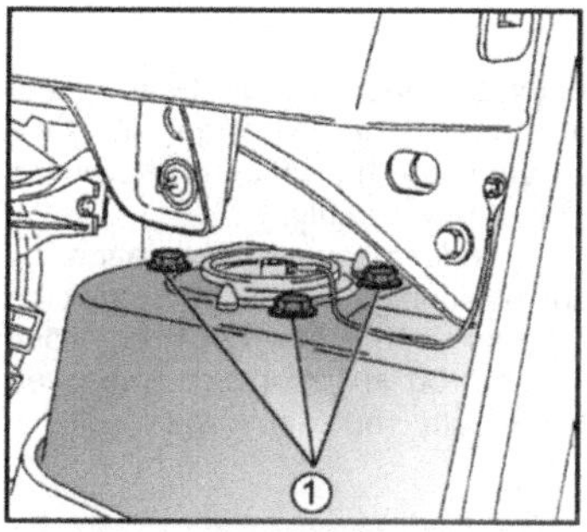

Fig. 5.2 – The three bolts (1) retain the upper spring strut bearing to the body panel.

- Apply the handbrake, engage 1st gear and slacken the wheel bolts.
- Place the front end of the vehicle on chassis stands and remove the front wheel. Place a mobile jack loosely underneath the front axle unit.
- Inside the vehicle, remove the lower part of the dashboard on the side in question. Remove the sound-proofing material and the earth lead and unscrew the three bolts in Fig. 5.2.

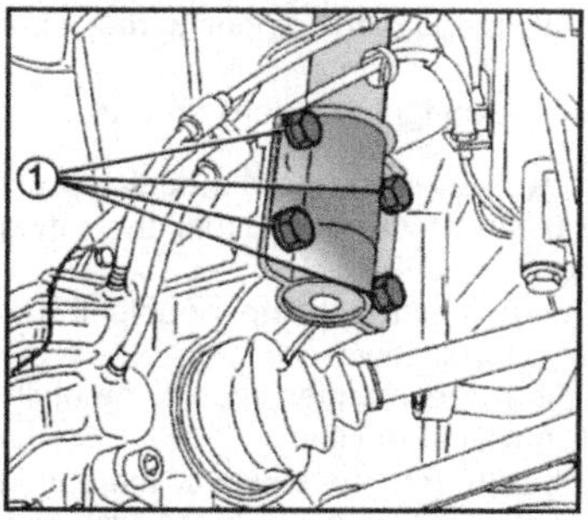

Fig. 5.3 – The four bolts (1) secure the spring strut to the steering knuckle. Steering knuckle, suspension arm, etc., remain in the vehicle when the spring strut is removed.

- Refer to Fig. 5.3 and remove the four bolts indicated by the arrows.
- Slowly lower the jack until the upper end of the spring strut is free of the body panel and release the strut from the steering knuckle. Withdraw the complete strut towards the rear. *Under no circumstances, remove the nut in the centre of the spring strut, as it would "explode".* If the spring strut is to be dismantled, you can slacken the nut but re-tighten it again before the strut is removed. The steering knuckle and the suspension arm remain in the vehicle.

5.2.1. DISMANTLING

Only the coil spring, the complete spring strut, the spring strut bearing and the sundry parts can be replaced. The spring strut cannot be dismantled. Never try to dismantle a spring strut, without compressing the spring first. Fig. 5.4 shows a compressor as used in the workshop. A similar tool, available from tool hire companies, for example, is suitable as long as the hooks can be placed over four or five spring coils.

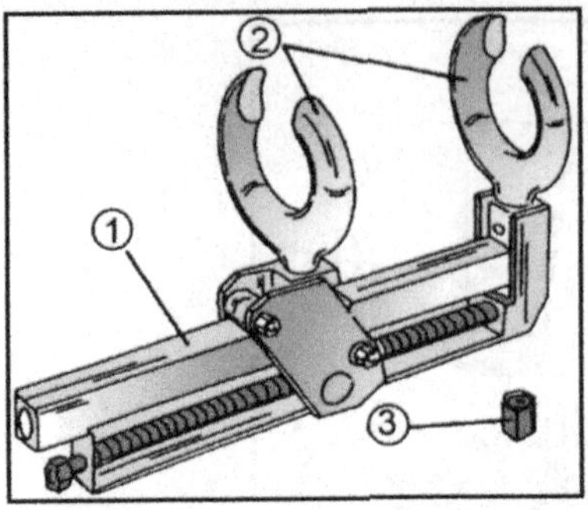

Fig. 5.4 – Special spring compressor for the dismantling of a front spring strut.
1 Compressor body
2 Claws
3 Nut

- Clamp the spring strut into a vice and fit the spring compressor. Fig. 5.5 shows the use of the spring compressor shown in Fig. 5.4. Make absolutely sure that the coils are securely retained in the compressor. Tighten the spring compressor until the spring can be rotated freely between the spring seats.
- Insert an Allen key into the end of the piston rod and slacken the piston rod nut (if not already slackened), using a flat or ring spanner of suitable size. Fully unscrew the nut and remove the upper spring strut bearing, the spacer, the needle thrust bearing, the upper spring seat and the coil spring. Remove the rubber rebound stop from the end of the piston rod and the strut from the compressed spring. The fitting order of the various parts is shown in Fig. 5.6.

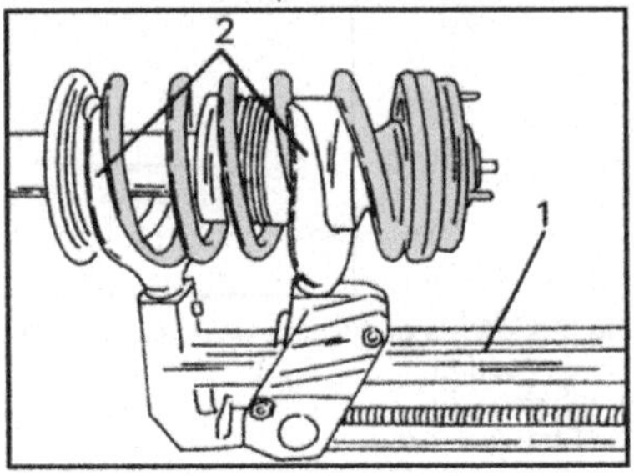

Fig. 5.5 – The spring compressor (1) can be clamped into a vice. The claws (2) are placed around the coils of the spring.

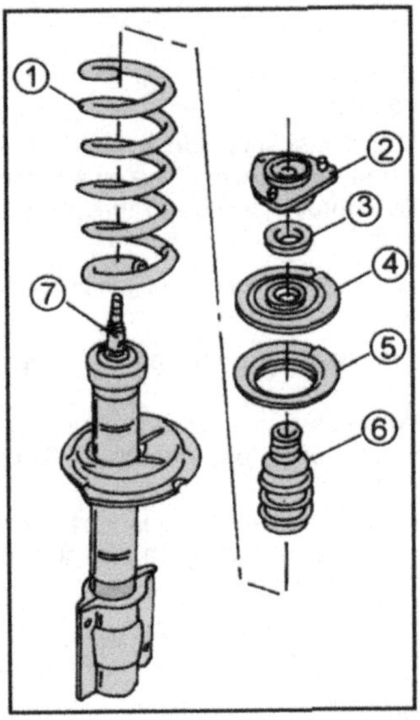

Fig. 5.6 – Dismantled spring strut.
1 Coil spring
2 Spring strut bearing
3 Thrust ring
4 Needle roller bearing
5 Upper spring seat
6 Rebound rubber
7 Spring strut

The rebound stop, the rubber protector, the needle thrust bearing and the spring strut bearing must be replaced if no longer in good condition. Take care not to damage the smooth surface of the piston rod when the spring strut is stored for later assembly.

The shock absorber can be checked for correct operation by clamping the spring strut in upright position into a vice. Screw the back onto the piston rod and grip the nut with a pair of pliers. Move the piston rod up and down. The piston rod should move with even resistance. Any "dead play" indicates a faulty shock absorber.

5.2.2. ASSEMBLY

Before assembly of the spring strut, prepare the upper spring strut bearing.

- Place the upper cap onto its edge (upright position) and, using a small chisel or a screwdriver, knock the needle thrust bearing (4) in Fig. 5.6 out of the cup. Note the fitting direction of the bearing.
- Grease the locating bore and press the needle thrust bearing into the spring cup as shown in Fig. 5.7. If the bearing is difficult to fit in the manner shown, carefully use a rubber or plastic mallet to tap it in.
- Insert the rebound rubber (6) into the inside of the spring cup and fit the remaining parts in accordance with Fig. 5.6.
- Clamp the spring strut in upright position into a vice and pull the piston rod out of the spring strut as far as it will go. Fit the compressed spring over the spring strut, engaging the lower coil with its stop. Place the upper spring seat over the spring.

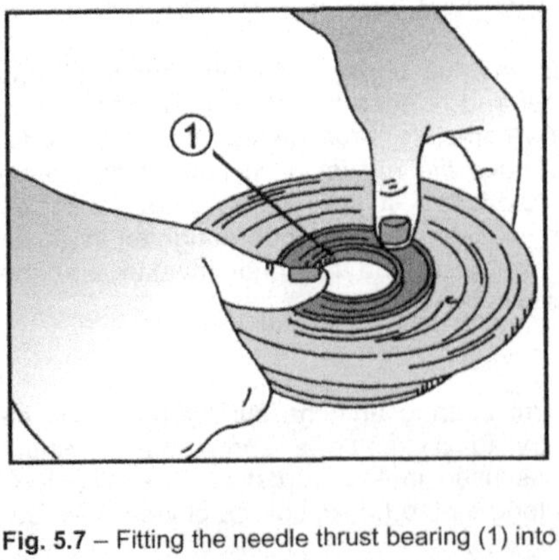

Fig. 5.7 – Fitting the needle thrust bearing (1) into the upper spring cup.

- Fit the assembled spring strut bearing over the piston rod, followed by the spacer washer.
- Fit a new self-locking nut to the end of the piston rod and tighten the nut to 7.0 kgm (50.5 ft.lb.). The piston rod must be held against rotation until it is possible to apply a socket with torque wrench to the nut. The latter should be tight enough to keep the piston rod in position.
- Slowly release the spring compressor, at the same time checking that the spring coils enter the locations at top and bottom. The ends of the coils must be resting against the stop on the spring strut and a similar stop on the upper end.

5.2.3. INSTALLATION

The installation of the spring strut is a reversal of the removal procedure. Note the following points.

- Insert the spring strut with the studs from below into the holes in the body panel and fit the three nuts without tightening them. Always use new self-locking nuts.
- Lift the complete front suspension unit with the jack and engage the spring strut. Fit the four bolts shown in Fig. 5.3. Tighten the bolts evenly across to 21 kgm (150 ft.lb.).
- Tighten the three nuts securing the upper spring strut bearing to 2.1 kgm (15 ft.lb.).
- Lower the vehicle to the ground and tighten the wheel bolts to the correct torque.

5.3. Removal and Installation of a Front Suspension Unit on One Side

The following instructions must be followed when the complete steering knuckle together with the drive shaft is to be removed. It should be noted that the inner mounting of the suspension arm must be tightened when the weight of the vehicle is resting on the wheels. This will pre-load the rubber mountings to their correct operating position. Fig. 5.8 shows the front suspension on one side.

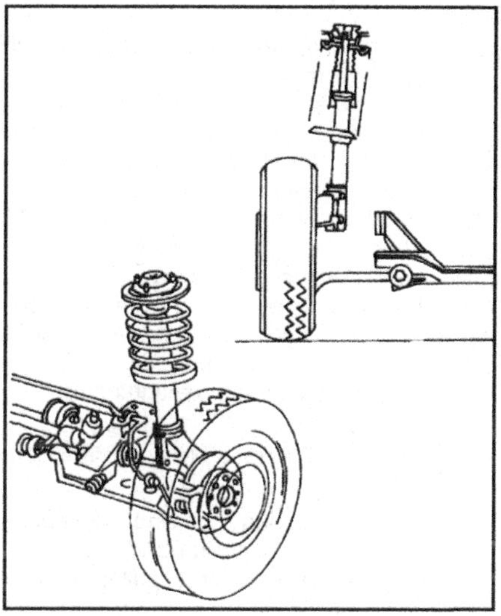

Fig. 5.8 – View of the front suspension on one side.

Attention: The hub grease cap is normally removed with a puller to be suitable for refitting. A damaged cap must be replaced.

- Slacken the wheel bolts and remove the hub grease cap with a strong screwdriver or a small chisel. It may be easier to remove the wheel as you will be able to insert the chisel straight into the gap. In this case, slacken the drive shaft nut before the wheel is removed. *Never attempt to slacken the nut when the vehicle is on chassis stands.*
- After slackening the drive shaft nut, place the front end of the vehicle on chassis stands and remove the wheel (as may be the case).
- Separate the connection between the brake hose and the metal pipe on the chassis. Plug the open ends of hoses/pipes in a suitable manner. Sticky tape is useful for this job.
- Disconnect the track rod ball joint from the steering lever and separate the connecting links for the stabiliser bar from the suspension strut.
- Slacken the bolts of the suspension arm mounting at the inside. In the engine compartment, remove the three nuts securing the upper spring strut bearing, as described during the removal of the spring strut. Support the suspension assembly from below, as it will drop as soon as the last nut has been removed.

- Withdraw the steering knuckle towards the outside, guiding the drive shaft. If the drive shaft is to be removed, follow the instructions in Section 4 for the side in question. Otherwise, withdraw the steering knuckle from the end of the drive shaft. Retain the drive shaft with a piece of wire as horizontally as possible.

The installation is a reversal of the removal procedure. Observe the tightening torques at the end of this section or refer to the instructions already given for the installation of the spring strut. Bleed the brake system after re-connecting the brake hose to the pipe. Fit the drive shaft as described in Section 4.

5.4. Dismantling and Assembling a Front Axle Unit

After the front axle unit has been removed in the manner described above, you will be able to dismantle it to replace individual parts. All tightening torques are given at the end of this chapter or have already been mentioned in earlier sections. Note the following.

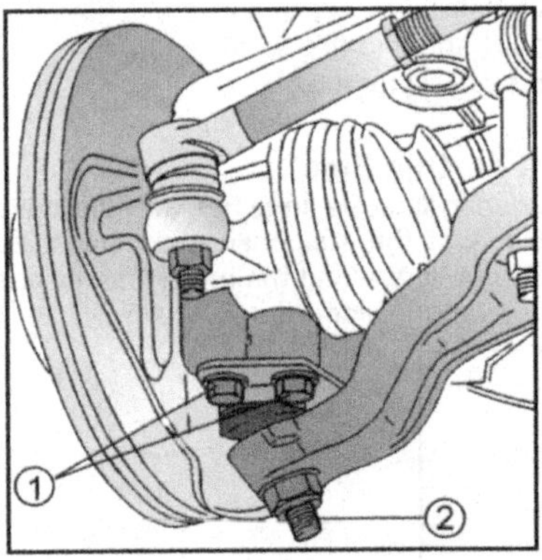

Fig. 5.9 – Attachment of the suspension ball joint to the steering knuckle.
1 Bolts, joint to steering knuckle
2 Nut, joint to suspension arm (note the tightening torque)

- If the suspension arm is to be removed together with the suspension ball joint, remove the bolts from the bottom of the steering knuckle. Otherwise, remove the ball joint nut from the bottom of the steering knuckle and separate the joint stud with a suitable puller. Fig. 5.9 shows the two possibilities.
- Remove the wheel bearings as described in the next section.

5.5. Steering Knuckle and Wheel Bearings

The replacement of the wheel bearings is not a straightforward operation, as a dial gauge with a suitable holder is required to adjust the wheel bearing clearance. Required are also a puller and press mandrels, which can, however, be substituted by make-shift or conventional tools.

The steering knuckle must be removed, i.e., either the spring strut or the complete front axle unit must be removed as already described. A wheel bearing can then be removed as follows.

- Remove the wheel hub from the steering knuckle. Alternative methods are available. Either you place the steering knuckle under a press and remove the hub from the rear, using a suitable press mandrel or you use a puller as shown in Fig. 5.10 and push out the wheel hub. The pressure of the centre spindle must be applied to the outer diameter of the hub, i.e., a suitable plate must be placed over the hub before the puller is applied. The spindle can then be tightened.

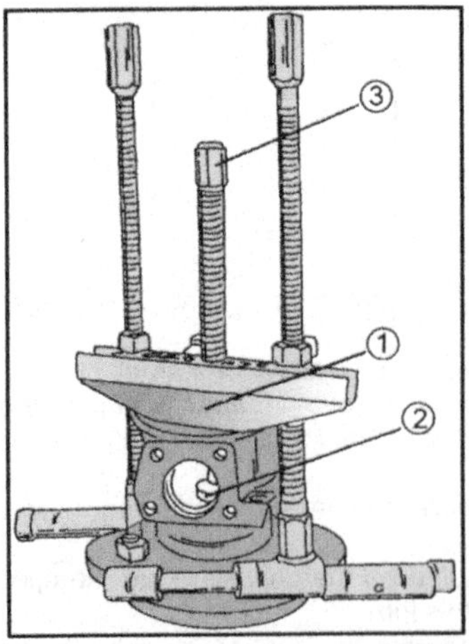

Fig. 5.10 – The wheel hub can be removed with a puller (1). Place thrust washer (2) over the wheel hub face. The spindle (3) is tightened to push out the hub.

- The bearing race and the bearing adjustment spacer will remain on the wheel hub and must be withdrawn with a suitable puller. The claws of the puller must be engaged under the bearing race.
- Place the steering knuckle over a vice and drive the two bearings out of the inside of the steering knuckle. The two grease seals will come away together with the bearing in question.
- Remove the two outer bearing races from opposite sides of the steering knuckle. Collect the bearing spacer and the adjusting shims.

Thoroughly clean all parts and check them systematically for wear. Grease seals must always be replaced, once removed. Discoloured bearings are also due for replacement.

The wheel bearings are refitted in the following manner. The description is based on the use of the special tools, but suitable alternative tools can be used as applicable.

- Make sure that the inside of the steering knuckle is clean and drive the outer bearing races from opposite sides into the steering knuckle, with the cone facing towards the outside. Take care not to damage the bearing running faces.

The bearing clearance (play) must now be adjusted. The work must be carried out with great care, as badly adjusted wheel bearings will soon lead to failure. The bearing clearance must be adjusted to 0.02–0.10 mm and must be measured with a dial gauge.

- Clean the inside of the filter outer bearing race, grease the bearing (6) in Fig. 5.11 and tap the bearing in position. The guide (2) and the assembly bolt are used for this purpose.
- Insert the bearing spacer (7) into the outside of the steering knuckle and the inner wheel bearing (8) from the other side. Insert the two thrust pieces (3) and (4) into the opening. Fit the nut (5) to the assembly bolt (1) and tighten the nut to 1.0 kgm (7.2 ft.lb.). The bolt head must be held to prevent it from rotating.

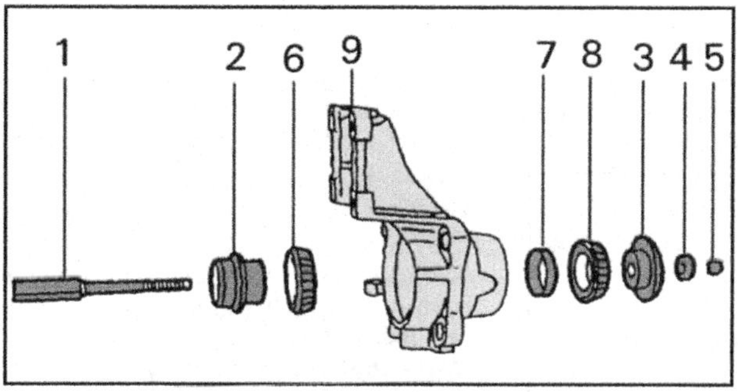

Fig. 5.11 – Fitting the wheel bearing to the steering knuckle, using the special tools.

1 Bolt
2 Thrust piece
3 Thrust piece
4 Washer
5 Nut
6 Inner wheel bearing
7 Bearing spacer
8 Outer wheel bearing
9 Steering knuckle

- Apply a ring spanner to the nut and rotate the wheel bearings a few times to settle them in position.

Assembly Note: If the special tools cannot be obtained (we would say in most cases), use a long bolt. Suitable sockets can be used in place of items (3) and (2). Large washers and a nut complete the make-shift arrangement.

The pre-load of the bearings must now be adjusted. Again, the description is based on the special tools, but a suitable dial gauge, arranged as shown in Fig. 5.12, will do.

Fig. 5.12 – Checking the wheel bearing clearance. Place the stylus of the dial gauge (3) on the end of the bolt (1). Moving the bolt up and down will indicate the clearance of the wheel bearings. The dial gauge holder (2) is attached to the steering knuckle.

- Clamp the steering knuckle into a vice as shown in Fig. 5.12, attach the dial gauge holder and place the stylus onto the end of the bolt as shown in the illustration. Set the dial gauge to "Zero".
- Grip the end of the bolt (1) and move it to and fro (up and down), at the same time observing the dial gauge reading. The indication must be between 0.02 and 0.10 mm. If the clearance is not within the limits shown, increase or decrease the height of the bearing spacer (7) in Fig. 5.11. Bearing spacers are available in thicknesses from 21.29 to 21.97 mm, in increments of 0.04 mm between sizes.
- Remove the used tools and take out the wheel bearing cones, of the bearing spacer must be replaced. In this case, re-check the end float as described.
- Generously grease the bearing (6) in Fig. 5.11 and insert it into the steering knuckle. Fit the grease seal into the steering knuckle until the outside is flush.

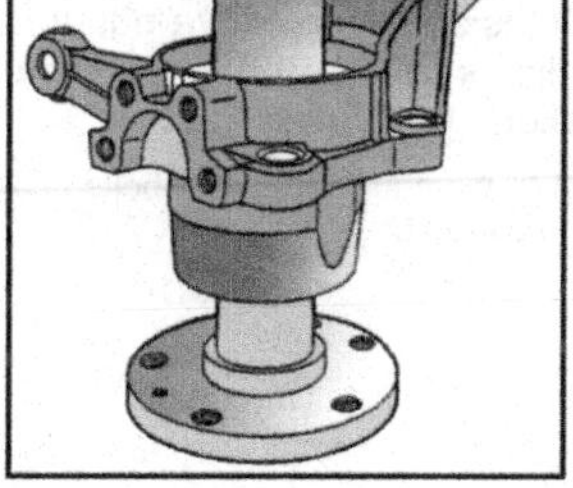

Fig. 5.13 – Fitting the wheel hub. Place the steering knuckle with the fitting bearing over the hub and press home.

- Place the wheel hub with the flange onto the press table and drive the steering knuckle and the inserted bearing over the hub. A piece of tube of suitable diameter must be used. The tube must be placed against the inner bearing race. Drive the bearing fully over the hub. The hub and steering knuckle must be supported, as shown in Fig. 5.13.
- Place the selected bearing spacer inside the steering knuckle.
- Fit the well-greased inner bearing race (6) in Fig. 5.11 into the steering

knuckle (at least 50-g wheel bearing grease must be used). Drive the bearing in position with a piece of tube of suitable diameter. The wheel hub/steering knuckle remains in the position shown in Fig. 5.13. Grease a new oil seal and carefully fit it into the steering knuckle until the end is flush the knuckle face.

5.6. Suspension Arms – Removal and Installation

The following text describes the removal of the suspension arms. The remaining parts of the front suspension are not removed.

Before the front end of the vehicle is placed on the chassis stand, it will be necessary to remove the stabiliser bar from the bottom of the suspension arms, to release the tension from the arm. We would like to point out that the space is very confined to remove the mountings when the wheel is on the ground. The nuts (1) and (2) in Fig. 5.14 must be removed to release the outer ends of the stabiliser bar from the suspension. The bar can be left on the chassis. Removal takes now place as follows.

- Slacken the wheel bolts, place the front end of the vehicle on chassis stands and remove the wheel.

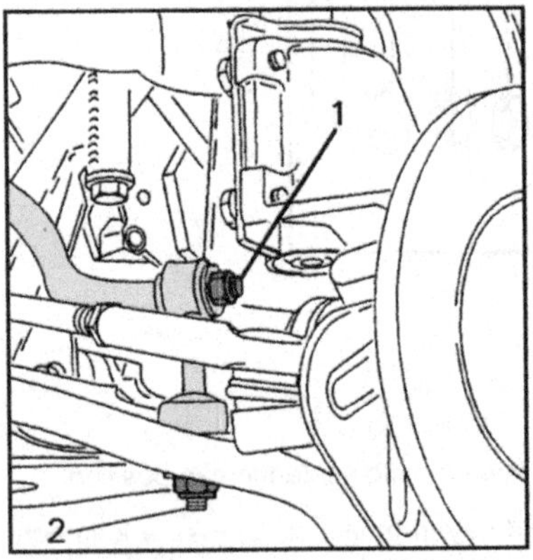

Fig. 5.14 – The stabiliser bar connecting linkages are secured with nuts at the upper and lower ends. The nuts (1) and (2) must always be replaced.

- Remove the protective shield underneath the vehicle.
- Refer to Fig. 5.9 and remove the ball joint nut (2). Separate the ball joint with a suitable puller from the suspension arm. ***Note that the tightening torque for the nut is not the same on all models.***
- On the inside of the suspension arm, remove the arm mounting bolts shown in Fig. 5.15. ***The three bolts have a different tightening torque.***
- Remove the suspension arm.

The installation is a reversal of the removal procedure, but different tightening torques must be observed. The suspension arm mounting bolts must be tightened when the wheels are back on the ground. Note the following.

- Place the rear end of the suspension arm against the sub-frame. Use a jack to push the arm well against its mounting points.
- Fit the bolt (1) in Fig. 5.15, followed by the bolts (2) and (3). Tighten all bolts finger-tight. After the vehicle is on its wheels, tighten the bolt (1) with 25 kgm (180 ft.lb.), the bolt (2) with 20 kgm (144 ft.lb.) and the bolt (3) with 17 kgm (122.5 ft.lb.).

Connect the lower ball joint to the steering knuckle and fit a new nut. The nut is tightened to 13 kgm (97.5 ft.lb.), with the exception of the "18Q" model. The nut on these vehicles is tightened to 23.5 kgm (170 ft.lb.). The nut can be tightened with the wheels off the ground.

- Refit the stabiliser bar, referring to Fig. 5.14. Both nuts are tightened to 8.0 kgm (58 ft.lb.) when the vehicle is with the wheels on the ground.
- All other operations are now carried out in reverse order. Lower the vehicle on its wheels and tighten the elements of the front suspension as specified above.

5.7. Suspension Ball Joint – Replacement

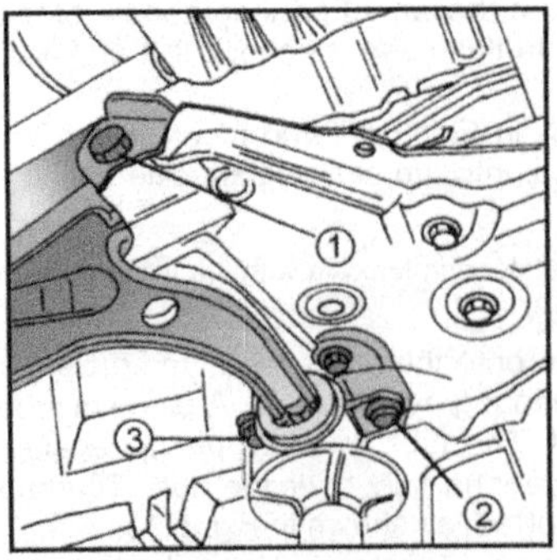

The replacement of the ball joint follows a similar pattern as described for the removal of the suspension arm. After the suspension arm has been separated from the ball joint (see Fig. 5.9), remove the two bolts (1). At the outside, a protective panel is fitted, as can be seen in Fig. 5.16, which is removed after removal of the bolt. The ball joint can now be taken off.

Fig. 5.15 – Attachment of a suspension arm to the sub-frame. The bolts (1)–(3) are tightened to the torque setting given in the text.

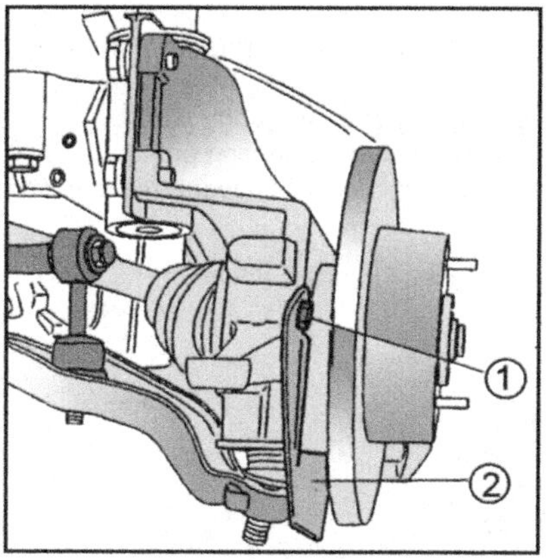

Fig. 5.16 – The protective panel (2) on the outside of the steering knuckle is secured by the screw (1).

The installation is a reversal of the removal procedure. Tighten the bolts (1) to 8.8 kgm (63.5 ft.lb.). Tighten the nut (2) as described during the installation of the suspension arm (Section 5.5.).

5.8. Stabiliser Bar – Removal and Installation

As the track rod ball joint must be separated from the steering lever, a ball joint puller will be necessary.

- Slacken the wheel bolts, place the front of the vehicle on chassis stands and remove the wheel. Also, remove the protective shield underneath the front end of the vehicle.
- Remove the track rod ball joint nut and separate the joint with a puller, as shown in Fig. 5.17. Move the track rod upwards to have it out of the way.

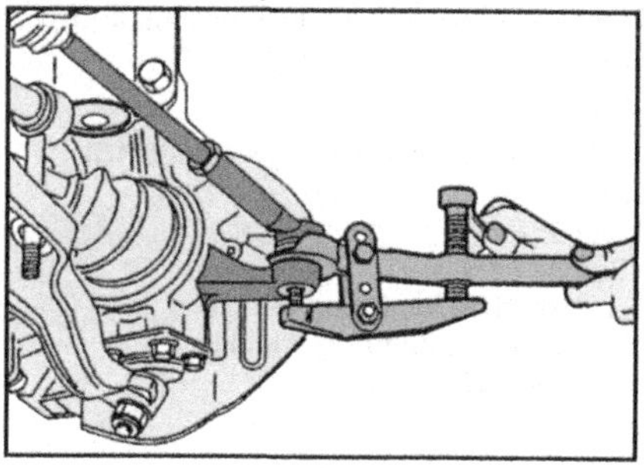

Fig. 5.17 – Separating a track rod ball joint.

- Refer to Fig. 5.18 and remove the nuts (2) and (3) securing the ends of the connecting links.
- From below the vehicle, remove the mounting clamp bolts on the L.H. and R.H. sides at the locations shown in Fig. 5.19 and remove the stabiliser bar from the R.H. side of the vehicle. The bar must be lifted and rotated towards the bottom to release it from the suspension assembly.

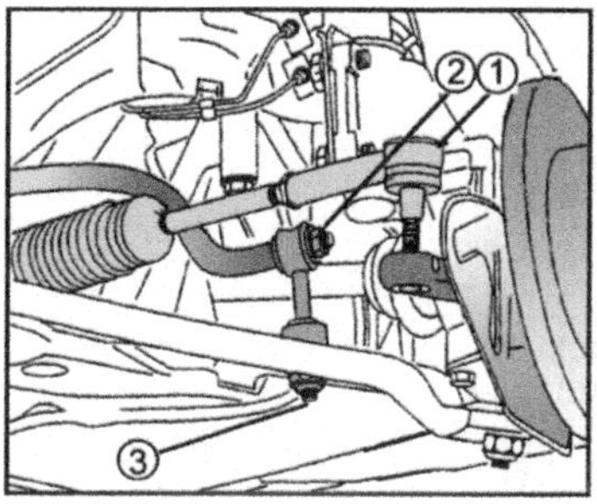

Fig. 5.18 – Stabiliser connecting linkage connected to the bar and the suspension arm. The track rod ball joint has already been separated.

1 Ball joint
2 Stabiliser end
3 Suspension arm end

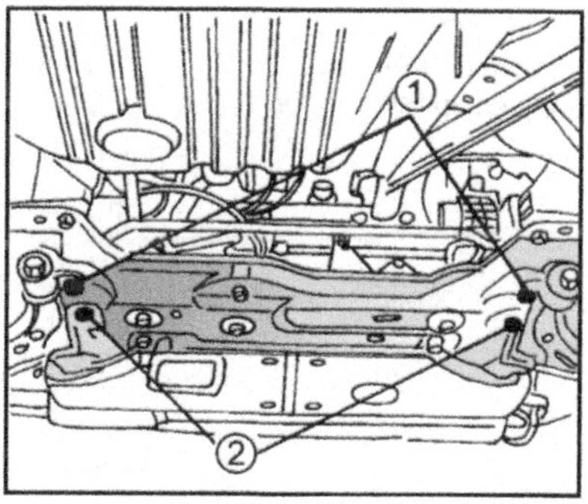

Fig. 5.19 – Bolts (1) and (2) secure the stabiliser bar mounting clamps to the bottom of the crossmember.

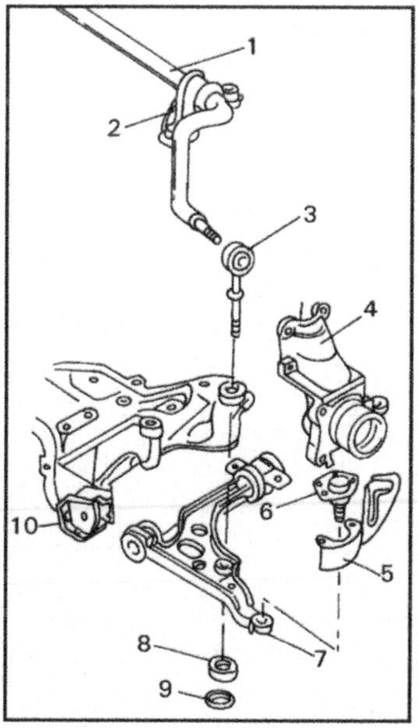

Fig. 5.20 – Stabiliser bar and lower suspension arm.

1 Stabiliser bar
2 Mounting clamp
3 Connecting rod
4 Steering knuckle
5 Protective cover
6 Suspension ball joint
7 Suspension arm
8 Rubber bush
9 Metal cup
10 Suspension crossmember

The installation is a reversal of the removal procedure. The self-locking nuts must be replaced. Bolts and nuts are tightened with the wheels back on the ground. Bolts (1) in Fig. 5.19 are tightened to 20 kgm (145 ft.lb.) and bolts (2) to 8.5 kgm (61 ft.lb.). Nuts (2) and (3) in Fig. 5.18 are tightened to 8.0 kgm (58 ft.lb.). Fig. 5.20 shows the attachment of the stabiliser bar together with other parts of the suspension.

5.9. Vehicles With Four-Wheel Drive

The front suspension is, in general, of the same design and construction, but the following differences should be noted.

- The upper spring strut mounting has 30-mm-thick inserts between spring strut and body panel.
- The 10-mm-thick inserts are fitted between the sub-frame and the body at the mounting points.
- The stabiliser bar, fitted as standard equipment, is pre-formed to avoid contact between the bar and the propeller shaft or the transfer box.

The repair operations described above are not influenced by the alterations specified above.

5.10. Tightening Torque Values

Ball joint to suspension arms:	
- Models 10Q and 14Q	13.0 kgm (94 ft.lb.)
- Model 18Q	24.0 kgm (173 ft.lb.)
Ball joint plate to steering knuckle	8.8 kgm (63.5 ft.lb.)
Nuts, stabiliser connecting links	8.0 kgm (57.5 ft.lb.)
Stabiliser bar to crossmember (Fig. 5.19):	
- Bolt 1	20.0 kgm (144 ft.lb.)
- Bolt 2	8.5 kgm (61.5 ft.lb.)
Suspension Arm to Chassis (Fig. 5.15):	
- Bolt 1	25.0 kgm (180 ft.lb.)
- Bolt 2	20.0 kgm (144 ft.lb.)
- Bolt 3	17.0 kgm (122.5 ft.lb.)
Track rod ball joint nuts	7.0 kgm (50.5 ft.lb.)
Spring strut to steering knuckle	21.0 kgm (151.5 ft.lb.)
Spring strut to body	2.1 kgm (15 ft.lb.)
Drive Shaft Nuts:	
- Models 10Q and 14Q	45.0 kgm (324 ft.lb.)
- Model 18Q	50.0 kgm (360 ft.lb.)
Brake caliper to spring strut	21.0 kgm (151.5 ft.lb.)
Brake caliper, guide bolts	2.7 kgm (19.5 ft.lb.)
Wheel Bolts:	
- Models 10Q and 14Q	16.0 kgm (115 ft.lb.)
- Models 18Q	18.0 kgm (130 ft.lb.)
Brake disc to hub	1.0 kgm (7.2 ft.lb.)
ABS wheel speed sensor	1.0 kgm (7.2 ft.lb.)

6. STEERING

6.0. Technical Data

Type	Rack and pinion steering, without or with power assistance. Mainly power-assisted steering
Turning circle between kerb stones/walls:	
- Short wheelbase (2.85 m)	11.00/11.50 m
- Medium wheelbase (3.20 m)	12.10/12.68 mm
- Long wheelbase (3.70 m)	13.70/14.14 m
Steering Pump:	
- Drive	Belt drive
- Fluid pressure	100 bar (1420 psi.)
Steering Fluid:	
- Filling capacity	1.3 litres
- Fluid type	ATF fluid, as in automatic transmissions
- Fluid level check	Every 6000 miles
Front Wheel Alignment (also refer to Section 5.0 for laden vehicle):	
- Castor, not adjustable	1° +/– 30'
- Camber, not adjustable	0° +/– 30'
- Wheel toe-in, adjustable	0° +/– 1 mm (0 +/– 0.04 in.)
- Checking conditions	Tyre pressures correct, vehicle empty

6.1. Maintenance Operations – Steering

Checking Steering Rack Gaiters: The steering rack is closed off at both ends with black rubber gaiters. These should be checked periodically for cuts or other damage to prevent water, dust or other foreign matter from entering the steering. To check the gaiters, turn the steering wheel into full lock and check the stretched gaiter on one side for cuts. Replace if a cut is visible. Turn the steering to the opposite lock and check the other gaiter. Make sure that the gaiter securing clamps are properly tightened.

Checking Rubber Dust Caps and Ball Joints: Check the rubber dust caps of each steering track rod ball joint for cuts. If cuts are detected, replace the joint as described later on.

The track rod ball joints can be checked for excessive wear. Jack up the front end of the vehicle and move the wheel to and fro, at the same time observing the track rod ball joint. Unusual movement of the joints can be seen. It is also possible to hold the ball joint where it enters the steering lever and ask a helper to turn the steering wheel to and fro (wheels on the ground). Movement in the joints requires the replacement of the joint(s).

Checking Steering for Play: Grip the steering wheel through the window and move the wheel to and fro, at the same time observing the front wheel. The road wheel must move immediately in response to the steering wheel movement. Otherwise: excessive play in the steering rack, worn track rod ball joints and worn universal joint on the steering shaft.

Checking the Fluid Level: See description later on.

6.2. Steering Unit

6.2.0. REMOVAL

Note: Before separating the steering shaft from the steering pinion, lock the steering wheel with the road wheels in the straight-ahead position. This will prevent damage to the cancelling switch underneath the steering wheel (use a piece of string in a suitable manner). **If an airbag is fitted, you will need some experience with that type of equipment. Otherwise, have the steering changed by a workshop.**

- Disconnect the battery negative cable. If an airbag is fitted, wait at least 1 minute before carrying out any other operation.

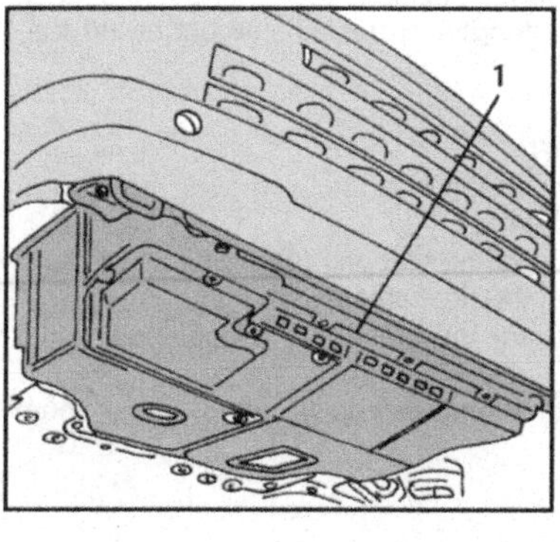

Fig. 6.1 – The protective shield (1) is fitted underneath the front end of the vehicle and can, however, have a different shape.

- Slacken the wheel bolts, place the front end of the vehicle on chassis stands and remove both wheels.
- Remove the protective shield underneath the front of the vehicle. Fig. 6.1 shows the shield in position, but different models have different arrangements. A second shield is fitted at the rear of the engine, which must also be removed.

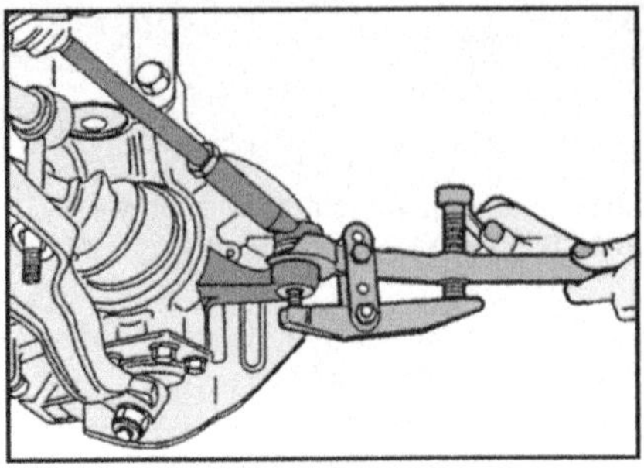

Fig. 6.2 – Separating a track rod ball joint.

- Turn the steering into full lock to one side and remove the nut securing the track rod ball joint to the steering lever. Use a suitable puller, as shown in Fig. 6.2, and separate the ball joint connection. Turn the steering wheel into the opposite lock and carry out the same operation on the other side.
- Slacken the stabiliser bar connecting rod nuts on both sides of the vehicle, i.e., slacken nuts (2) and (3) in Fig. 6.3.
- From the inside of the engine compartment, remove the nut from the end of the clamp bolt. The bolt secures the upper part of the steering column universal joint. Mark the relationship between universal joint and steering shaft with paint.
- Disconnect the stabiliser bar connecting rods from the suspension arms (both sides).

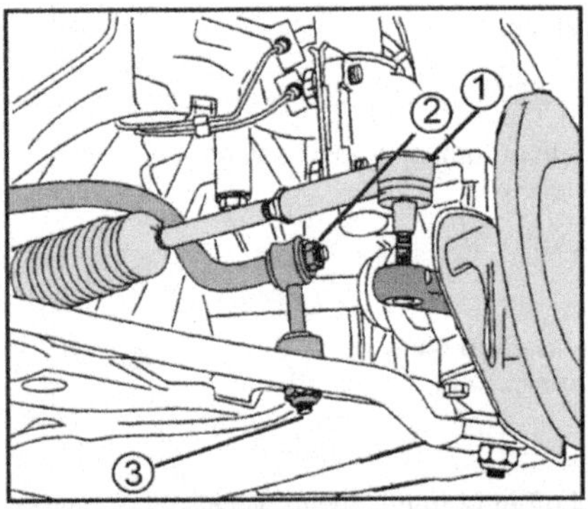

Fig. 6.3 – Details for the removal and installation of the steering.
1 Track rod
2 Nut, linkage to bar
3 Nut, linkage to suspension arm

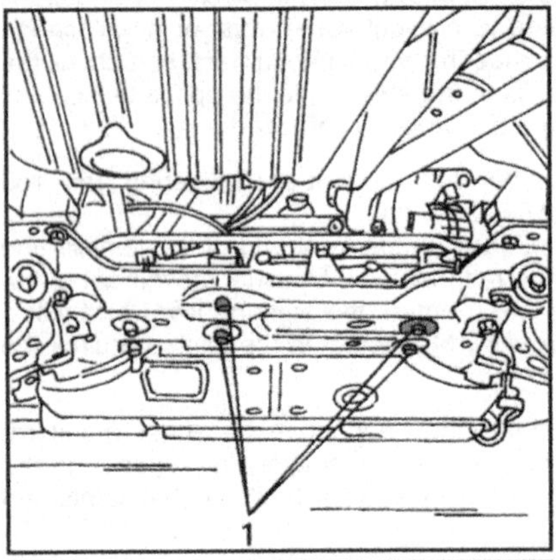

Fig. 6.4 – The four bolts (1) secure the steering unit to the front crossmember.

- From the bottom of the suspension crossmember, remove the four bolts (1) in Fig. 6.4.
- Using a tyre lever, push the stabiliser bar as far as possible towards the top and turn the steering until the steering universal joint can be disengaged from the steering pinion. Remove the steering sideways out of the vehicle (to the left or right). The removed steering is shown in Fig. 6.6.

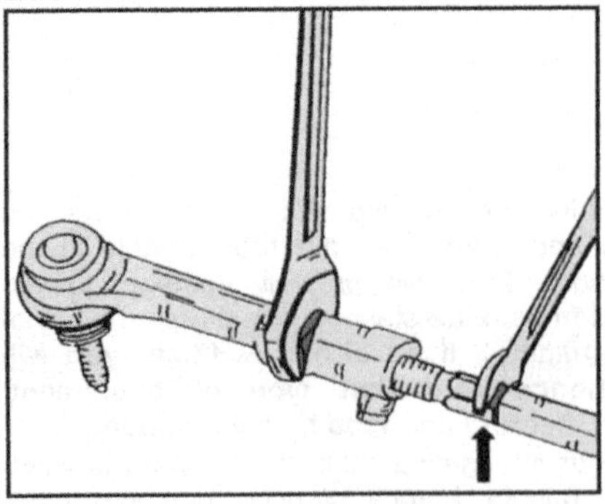

Fig. 6.5 – Removal of a track rod ball joint from the track rod. Apply an open-ended spanner at the position shown by the arrow.

- If the track rod ball joints are to be replaced, clamp the track rod into a vice, as shown in Fig. 6.5, slacken the locknuts and unscrew the track rod ends, counting the number of turns and if applicable half-turns necessary to fully unscrew a joint. Write the value down for each side for later fitting of the track rod end. If the new joints are fitted in the same position, they will set the track rods to the correct length for the adjustment of the toe-in.

6.2.1. INSTALLATION

The installation is a reversal of the removal procedure, noting the following points.

- Check the condition of the steering rubber boots (gaiters) and make sure that the boot clamps are tight.
- The securing bolts for the steering are tightened to 6.5 kgm (47 ft.lb.).
- If new track rod ball joints are to be fitted, screw them onto the end of the rod(s). Provisionally tighten the locknuts.

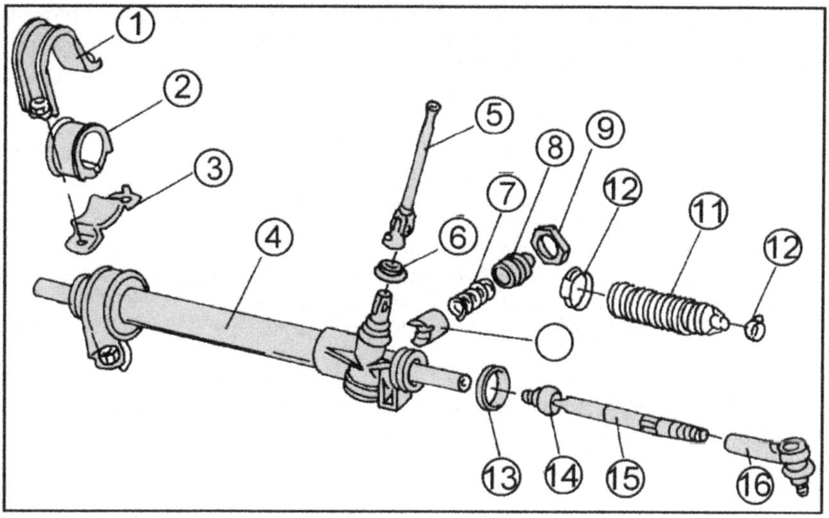

Fig. 6.6 – View of the removed and partially dismantled steering unit.

1 Mounting clamp
2 Rubber bush
3 Threaded holder
4 Rack housing
5 Steering shaft
6 Sealing grommet
7 Spring
8 Threaded sleeve
9 Locknut
10 Plunger
11 Rubber gaiter
12 Gaiter clamp
13 Sealing ring
14 Inner ball joint
15 Track rod
16 Track rod ball joint

- Clean the ball joint studs of oil or grease and connect with the steering levers. Tighten the nuts to 8.8 kgm (63 ft.lb.).
- Set the steering to the central position and connect the steering universal joint to the steering column shaft. Make sure that the marks made before removal are in line, insert the bolt and tighten the nut to 4.5 kgm (32 ft.lb.).
- Check the toe-in setting of the front wheels as described in Section 6.5.

6.2.2. STEERING REPAIRS

The steering should not be dismantled. In the case of wear or damage, fit a new steering or try to obtain an exchange steering. Steering with excessive play can, however, be adjusted. For safety reasons, however, we recommend to take the removed steering to a workshop where it can be checked and, if necessary, adjusted professionally.

The rubber gaiters can be replaced after disconnecting the track rods. You will have to assess how long the gaiter is damaged, as dirt may have entered the steering housing/rack. Before fitting a new gaiter, turn the steering wheel from lock to lock and clean off the old grease as good as possible. Then, re-grease the steering rack and fit the gaiter.

The inner ball joints at the ends of the track rods can also be replaced. Proceed as follows.

- Remove the rubber gaiter from the track rod and the side of the rack housing. Cut the old clamp band. On the other side, roll back the two rubber rings securing the gaiter to the track rod. Move the gaiter towards the outside of the track rod.
- Clamp the steering box into a vice (soft-metal jaws) and unscrew the ball joint housing from the steering rack. Workshops use a special wrench for this operation; you will have to use a pair of pliers or grips. The round-shaped ball joint housing is fairly sturdy as it is to be replaced; damage can be ignored. Unscrew the track rod as soon as the ball joint housing is free.

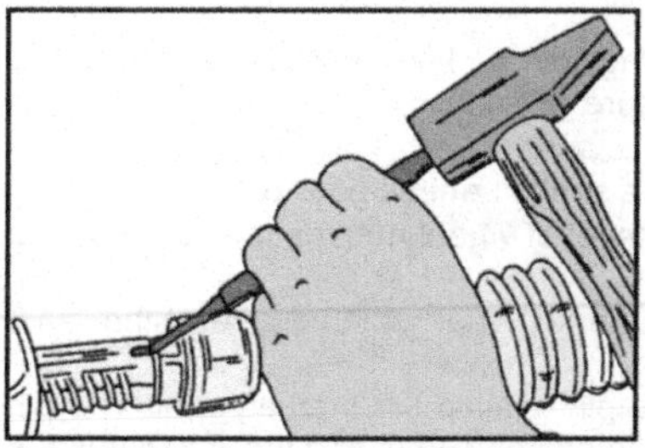

Fig. 6.7 – Locking a ball joint housing to the steering rack.

- Screw the new track rod with the new ball joint housing to the steering rack and tighten the housing with the grips as tight as possible.
- Secure the ball joint housing to the steering rack. The rack housing is clamped into a vice, and the lug on the ball joint housing is peened into the groove of the rack, as shown in Fig. 6.7, without splitting the metal.
- Grease the steering rack and refit the rubber gaiter.

- Fit the track rod ball joint the same number of turns. Tighten the track rod locknut provisionally. The final tightening takes place after the toe-in has been checked and/or adjusted – **do not forget to tighten the locknut after you have adjusted the toe-in.**

6.3. Steering Centre Position

The steering must be in the centre position during checks and adjustment of the toe-in or other operations, for example, during the removal and installation of the steering gear. Proceed as follows.

- Turn the steering wheel fully into one lock and attach a piece of sticky tape as the top of the steering wheel rim. Mark a line with a pencil into the tape.
- Turn the steering wheel into the opposite lock, counting the exact number of rotations.
- Divide the counted turns by 2 and turn the steering wheel towards the centre. The steering wheel must be in the centre position. If this is not the case, adjust the track rod on the side in question to set the wheel to the centre position. Check the position of the steering wheel and re-set, if necessary.

6.4. Front Wheel Geometry

The front wheel geometry must be checked when the vehicle has its kerb weight, i.e., unladen with coolant, fuel and oil.

6.4.0. CASTER ANGLE

The caster angle cannot be adjusted. The caster angle can be measured with conventional equipment, but an optical measurement by a specialist is preferred. If the obtained values are outside the limits, check the front suspension for distortion or better still, have it checked professionally. The same applies of the difference between the two sides exceeds 1°. Fig. 6.8 shows a diagram, showing the caster and camber settings. The caster diagram is demonstrated on the left.

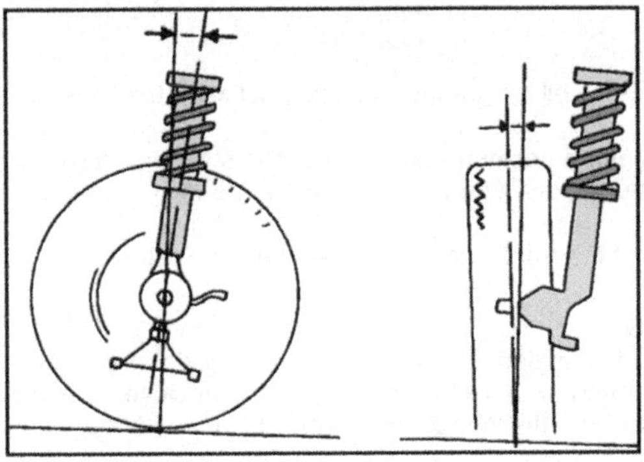

Fig. 6.8 – Diagram showing the caster angle (left) and the camber angle (right).

6.4.1. CAMBER ANGLE

The camber angle should be checked optically, but cannot be adjusted. If a mechanical camber gauge is used, follow the instructions of the manufacturer. The difference between the two sides must not exceed 1°. Incorrect camber angles are mainly due to distortion in the front suspension. Have the suspension checked, as recommended above. Fig. 6.8 demonstrates the camber angle on the R.H. side.

6.4.2. TOE-IN ADJUSTMENT

The front wheel toe-in can be measured with a mechanical tracking gauge. The front wheels are set to 0 mm, but with a tolerance of 1 mm towards toe-in or toe-out. Measure as follows.

- Check the tyre pressures and correct as necessary.
- Place the vehicle on level ground and turn the front wheels in the straight-ahead position.
- Place the tracking gauge with the pins against the wheel rims (level of wheel hubs) and set the gauge to "Zero". Mark the contact points with chalk or a spot of paint.
- Push the vehicle forward by half a turn of the wheels, i.e., the marked spot must now be at the read, again level with the wheel hubs.
- Push the tracking gauge underneath the vehicle and place one pin against the marked spot. The pin on the other side will most probably not reach the rim. Unlock the moveable pin and pull it towards the outside until it can be placed against the rim. Lock the pin in position.
- Remove the tracking gauge and read off the measurement. The indication should be "0" (Zero) or within a tolerance of 1 mm more or less.

- If an adjustment is necessary, slacken the locknuts at the inside of the two track rod ball joints and turn the track rod by the same amount on both sides. Note that there must be a gap of at least 2.0 mm (0.08 in) between the track rod ball joint and the locknut face. Make sure that the rubber gaiters cannot twist when the rods are turned. Slacken and then re-tighten the gaiter clamp bands, if necessary.
- Finally, tighten the two locknuts to 6.0 kgm (43.5 ft.lb.). The track rods must be held against rotation.

6.5. Power-Assisted Steering

6.5.0. REMOVAL AND INSTALLATION

Fig. 6.9 shows the details of the steering. A ball joint puller must be available to remove the steering. It is also advisable to have two clamps to pinch together the steering fluid hoses to prevent the draining of the complete system.

The removal and installation is carried out in a similar manner as described for the mechanical steering, with the difference that the pipes and hoses must be disconnected from the steering housing. Note the modification below.

Fill the steering system after installation as follows.

- Fill the fluid reservoir to the "Max" mark with the recommended steering fluid (Dexron II, ATF fluid). The 1.3 litres will be necessary if the system has been completely drained.
- Move the steering wheel a few times from one lock to the other (vehicle on chassis stands), with the engine switched off. Correct the fluid level, if necessary.
- Lower the vehicle to the ground, start the engine and again turn the steering wheel a few times from one lock to the other, until the fluid level in the reservoir remains steady. Refit the reservoir cap.

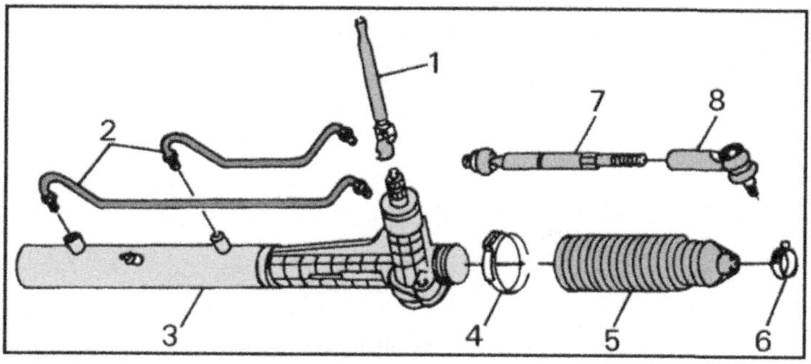

Fig. 6.9 – Component parts of the power-assisted steering.

1	Lower steering shaft	5	Steering gaiter
2	Fluid pipes	6	Outer gaiter clamps
3	Rack housing	7	Track rod
4	Inner gaiter clamp	8	Track rod ball joint

- After installation of the steering, check the toe-in setting or have it checked professionally.

6.5.1. STEERING FLUID LEVEL

The steering fluid reservoir has "Max" and "Min" marks. For every 6000 miles, the fluid level should be checked as part of a regular maintenance. Unscrew the cap to check the level. Never overfill the system. Make sure that the correct fluid is used to top-up the system.

If the fluid level has dropped below the minimum mark, bleed the system as described under "Removal and Installation of the Power-Assisted Steering" in Section 6.5.0.

To replace the fluid, disconnect the hose from the reservoir and let the fluid drain into a suitable container. Move the steering wheel to and fro to eject the fluid. Fill the system with 1.3-litre DEXRON II ATF fluid as described in Section 1.6.0.

Note – Modification of Steering Attachment

The steering attachment on models with the 2.0-litre engine and models with the remaining engines since approx. January 2000 have been modified. The steering securing screws have been changed from M12 thread to M14 thread. If a steering is replaced, you will have to give the model year (and the type of engine) in order to obtain the correct bolts.

The threads of M14 bolts must be coated with "Loctite". They have a spring washer and a plain washer underneath the bolt heads and are tightened to 11.0 kgm (79 ft.lb.). The inner diameter of the spacer sleeves for the bolts has been changed accordingly.

6.6. Tightening Torque Values

Steering securing bolts:	
- M12 bolts:	6.5 kgm (47 ft.lb.)
- M14 bolts:	11.0 kgm (79 ft.lb.)
Track rod ball joint locknuts	6.0 kgm (43.5 ft.lb.)
Track ball joints to steering levers	7.0 kgm (50.5 ft.lb.)
Locknuts at inside of track rods	12.0 kgm (86.5 ft.lb.)
Wheel bolts:	
- Models 10Q and 18Q	16.0 kgm (115 ft.lb.)
- Model 18Q	18.0 kgm (130 ft.lb.)
Steering universal joint clamp bolts/nut	4.5 kgm (32.5 ft.lb.)
Nuts for stabiliser bar connecting linkages	8.0 kgm (58.0 ft.lb.)
Hydraulic pipe connections	3.0 kgm (22 ft.lb.)
Fluid pipe clamp	0.9 kgm (6.5 ft.lb.)
Steering wheel nut	5.0 kgm (36.0 ft.lb.)

7. REAR AXLE AND SUSPENSION

The rear suspension of most models consists of two leaf springs, fitted in the longitudinal direction to the chassis side members. Each spring consists of a single spring leave. It is obvious that due to different load capacities of the various makes and models, there are differences in the carrying capacity of the springs. Whenever a spring is replaced, quote the model and the chassis number, i.e., 10Q, 14Q or 18Q. The springs are marked at the shackle mountings, at the rear ends of the springs.

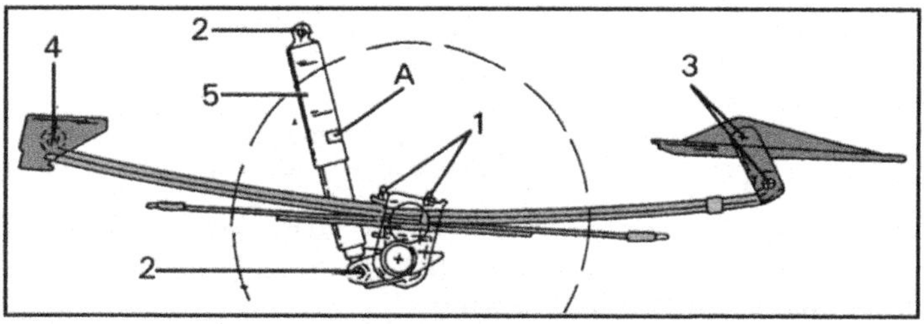

Fig. 7.1 – View of the rear suspension from one side.

1 Spring "U" bot Nut
2 Shock absorber bolts, 18 kgm
3 Spring shackle nut
4 Spring pin nut
5 Shock absorber
A Shock absorber marking

The springs are secured at the centre by two "U" bolts to the rear axle tube. A spring pin with a detachable bracket secures each spring at the front. A spring shackle arrangement with welded-in bolts and detachable shackles is used to attach the springs at the rear. Spring pins and spring shackle pins are used.

Two-way acting telescopic shock absorbers complete the rear suspension. The lower mounting points are located below the spring. Springs and shock absorbers are generally the same for vehicles with two-wheel and four-wheel drives. Figs. 7.1 and 7.2 show views of the rear suspension from different angles.

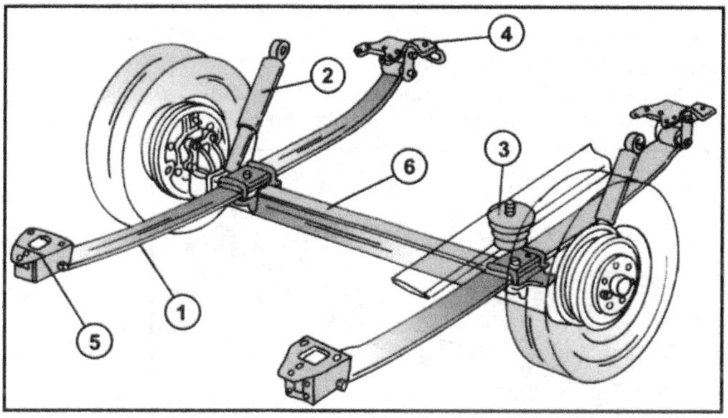

Fig. 7.2 – View of the assembled rear suspension together with the rear springs. The arrows show where the colour code can be found. Models with 4 × 4 drive have a packing underneath the rebound rubber.

1 Rear spring
2 Shock absorber
3 Rebound rubber
4 Spring shackle mounting bracket
5 Spring eye mounting bracket
6 Rear axle

From model year 2000 onwards, models 14Q and 18Q are fitted with a new rear suspension, consisting of coil springs, trailing arms and shock absorbers. The coil springs are not the same for all models. This type of rear suspension is shown in Fig. 7.3. The suspension is covered later on.

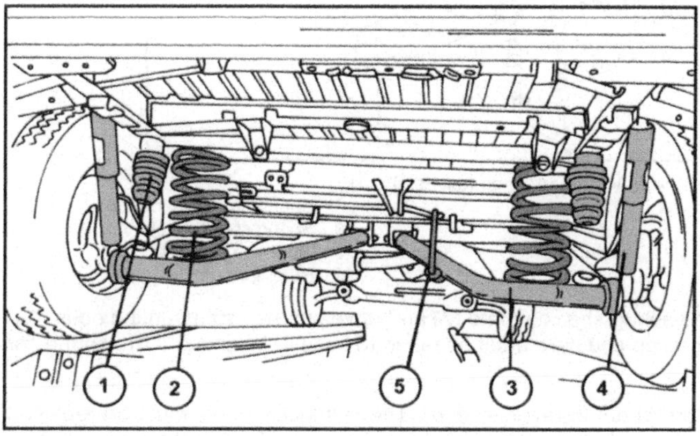

Fig. 7.3 – View of the rear suspension with coil springs.

1 Rebound rubber
2 Coil spring
3 Rear suspension arm (trailing arm)
4 Shock absorber
5 Brake pressure regulator linkage

7.0. Technical Data

Type	Leaf spring, one or two leaves and two-way acting telescopic shock absorbers or independent coil spring rear suspension.
Spring identification:	Colour-marking, depending on load capacity. Only fit springs with the same colour-marking if a spring is replaced. Different depending on model
Rear wheel hubs/bearings	Two taper roller bearings
Bearing end float	0.025–0.10 mm
Shock Absorbers:	
- Type	Differently depending on the Model
- Colour code	Black, blue or red, depending on the model
Tracking of rear wheels	0 +/– 1 mm (0 +/– 0.04 in.), not adjustable

7.1. Removal and Installation of a Rear Spring

- It should be noted that rear springs must be set to a certain height after installation. The necessary operations must, therefore, be followed.
- Remove the spare wheel to improve the access.
- Slacken the wheel bolts and jack up the rear end of the vehicle. Place chassis stands underneath the sides of the body.
- On the side where the spring is to be removed, place the jack underneath the axle tube and lift the axle until just under tension.
- Remove the nuts (3) shown in Fig. 7.4 from the spring "U" bolts (5) and carefully drive out the bolts towards the top. Take care not to damage the threads. To prevent thread damage, tap each side of the bolts alternatively until the "U" bolt is free. Remove the second "U" bolt in the same manner.
- At the rear of the spring, unscrew the two nuts (6) securing the spring shackle, and at the front of the spring, remove the nut at the inside of the spring pin (4). Do not remove the bolts at this stage.
- Remove the lower shock absorber mounting below the rear axle.
- Lower the rear axle a few inches on the jack and drive out the spring pin at the front. Take care that no parts of the rear axle are under tension during this operation (for example, the brake hose).

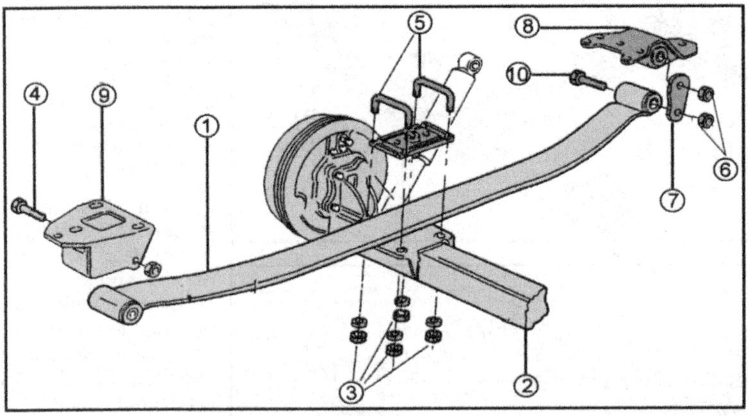

Fig. 7.4 – Details for the removal and installation of the rear springs (except 4 × 4).

1 Leaf spring
2 Rear axle tube
3 "U" bolt nuts
4 Spring pin
5 "U" bolts
6 Nuts
7 Spring shackle mounting bracket
8 Spring shackle bolt
9 Mounting bracket
10 Rear spring bolt

- At the rear of the spring, take off the spring shackle plate (7) and drive out the two mounting bolts (10). The spring can drop during the operation and care must be taken to keep it "balanced". The spring can now be removed.

The installation of the rear spring is carried out in reverse order. The self-locking nuts must be replaced. The spring must be set to a specific height. The mountings of the spring at the front and the rear and the shock absorber bolt must be tightened when the correct installation height has been obtained.

- Lift the spring into the correct position and fit the spring pin (4) at the front of the spring and the bolt (10) into the spring shackle at the rear end. Fit the nuts but do not tighten them (only hand-tight).
- Lift the axle on the jack until it rests against the spring. The centre bolt in the spring must engage into the hole in the axle pad.
- Drive the "U" bolts (5) from the top into the axle. If they cannot be inserted with a slight tap, check them for distortion. Distorted bolts can be straightened in a vice. Place a piece of tube over the legs of the bolt and bend them inwards or outwards as applicable. Screw on the nuts (3) and tighten them evenly to 13 kgm (94 ft.lb.).
- Connect the shock absorber at the bottom of the axle.
- The axle must now be lifted to bring it to the correct position in relation to the spring. To do this, lift the axle until dimension "H" in Fig. 7.5 is 192 mm in the case of models 10Q and 14Q or 297 mm in the case of model 18Q. A metric ruler should be used for the check.

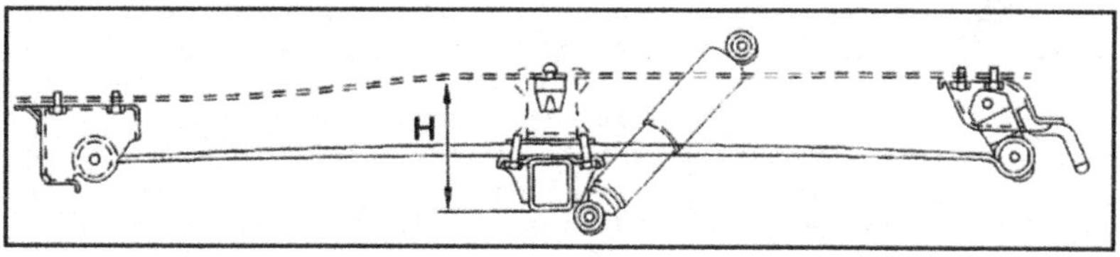

Fig. 7.5. – Measure the height "H" before tightening the rear spring and shock absorber bolts are tightened. Either lift the axle or place weights in the cargo room, depending on which operations are being carried out.

- Leave the axle and spring in this position and tighten the bolts (4) and (10) in Fig. 7.4 with 15.5 kgm (112 ft.lb.). Tighten the shock absorber bolt to 16 kgm (115 ft.lb.).
- Refit the wheel, lower the vehicle to the ground and tighten the wheel bolts to the specified torque.

7.2. Removal and Installation of the Rear Axle

- Slacken the wheel bolts, jack up the rear of the vehicle and place chassis stands underneath the sides of the body. The axle is removed without rear wheels.
- Follow the routing of the brake pipes and separate the connections between the brake pipe and the brake hose on the frame. First, unscrew the union nut and pull out the pipe end. Immediately protect the open end in a suitable manner (sticky tape). Knock out the spring plate and remove the hose fitting out of the metal bracket. Again, protect the open hose end by closing it off. Fig. 7.6 shows the brake hose/pipe connection.

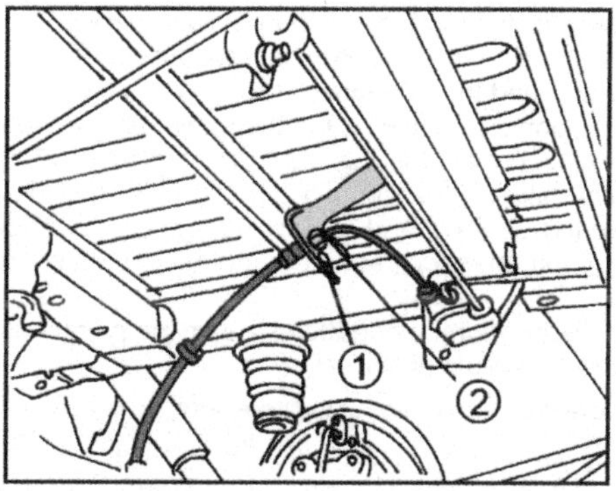

Fig. 7.6 – Separate the brake hose/brake pipe connections at (2) and drive out the spring plate (1).

- If ABS is fitted, separate the hydraulic pipe connections, drive out the clips securing the brake hose, disconnect the plugs for the wheel speed sensors on the cable harness and free the cable harness from its attachment on the body. Plug open connections to prevent the entry of dirt. Fig. 7.7 shows where parts must be disconnected.
- Remove the nut from the end of the handbrake cable on the brake compensation lever and disengage the cable end. The outer sleeves of the handbrake cables must be detached from the bracket on the frame and pulled out.

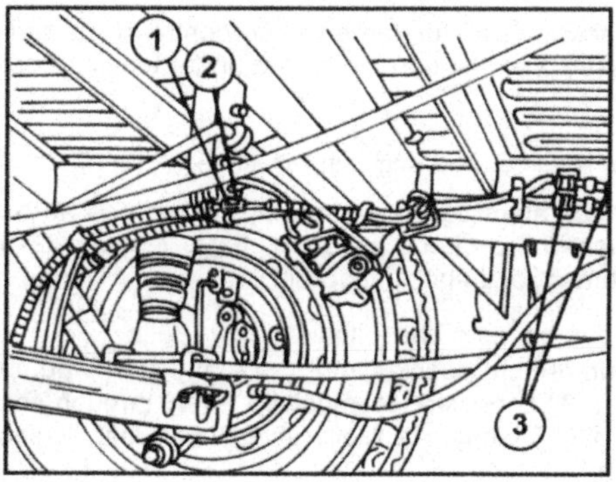

Fig. 7.7 – View from underneath the vehicle when ABS is fitted.

1 Spring clip for hose attachment
2 Brake hose connection
3 Connector for wheel speed sensor

- Remove the spring clip from the brake pressure regulator linkage at the position shown in Fig. 7.8 and disengage the linkage. Push the linkage to one side where it cannot be in the way.

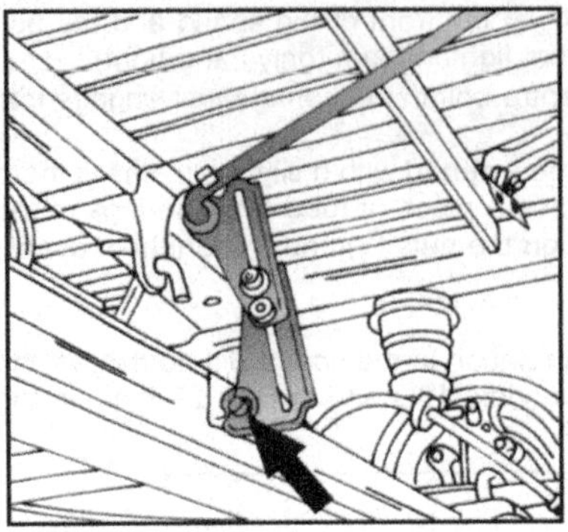

Fig. 7.8 – The position of the clip securing the brake pressure regulator linkage.

- Place a jack underneath the centre of the rear axle and lift the axle until the shock absorbers are compressed. Remove the upper and lower shock absorber mountings and lift out the two units. Another possibility is to remove the shockers from their upper mountings and leave them attached at the bottom.
- Check that the axle is securely supported on the jack and remove the bolts securing the spring pin bracket at the front of the spring and the bolts securing the spring shackle to the chassis frame at the rear. The 12 bolts must be removed in total. There is no need to remove the nuts securing the spring pins and the spring shackle plates. The spring eyebolts at the front can, however, be slackened if it is intended to remove them later on.
- Undo the exhaust system mounting to make sure it cannot be in the way during the removal of the axle.
- Lower the rear axle on the jack until it can be removed towards the rear. A helper should steady the axle on the jack. Make sure that none of the removed items can get caught in the axle.

The installation of the rear axle is made easier when the spring pin and the shackle plate nuts are slackened before the axle is lifted in position. Self-locking nuts must be replaced.

- Roll the axle underneath the vehicle and lift it in position until the rear mounting bracket is snug against the mounting face on the chassis. Fit the mounting bolts without tightening them fully.
- The mounting bracket must now be centred. The workshop used special centering pins, which are inserted, into one of the mounting holes. As these will not be available, insert a drift of suitable diameter. Fit the bracket securing bolts and tighten them to 14 kgm (101 ft.lb.).
- Lift the axle further, as required, until the front spring mounting bracket is against the chassis. Fit the bolts without tightening them fully. Check that all bolts are situated in the centre of the elongated holes and then tighten them to 14 kgm (101 ft.lb.).
- Re-connect the shock absorbers without tightening the bolts and nuts.
- Insert the brake hose through the hole in the bracket and secure it with the spring plate. Re-connect the brake pipe to the brake hose and tighten the union nut. Take care not to twist the hose. Also, connect the linkage for the brake pressure regulator and the handbrake cable.
- The rear suspension must now be loaded down as specified during the installation of the rear springs in Section 7.1. Important is dimension "H" in Fig. 7.5.
- Tighten the spring shackle and the spring eye pin to 15.5 kgm (112 ft.lb.) and the shock absorber mounting nuts to 16 kgm (115 ft.lb.).
- Finally, adjust the handbrake and bleed the brake system as described in the "Brakes" section.

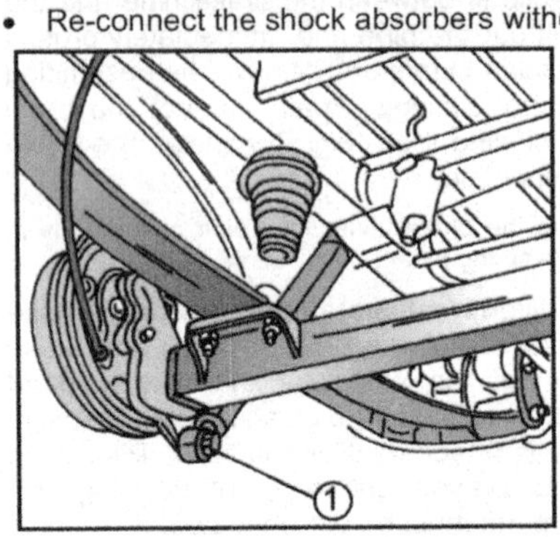

Fig. 7.9 – The shock absorber mounting at the lower end.

7.3. Dampers – Removal and Installation

- Place the rear of the vehicle on chassis stands and place the jack underneath the centre of the rear axle. Lift the axle until the shock absorbers are compressed, i.e., they are no longer under tension from the springs.

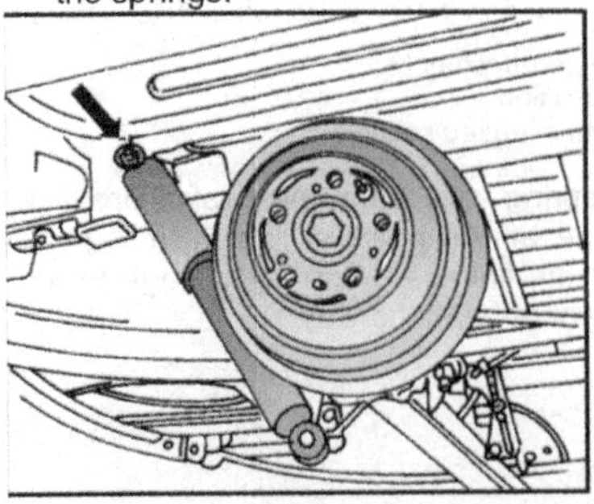

Fig. 7.10 – The shock absorber mounting at the upper end.

- Remove the shock absorber mountings at the top and at the bottom and remove the shock absorber. Figs. 7.9 and 7.10 show the attachment of a shock absorber.

A shock absorber can be tested as follows.

- Clamp the unit in upright position into a vice and slowly pump it up and down. The resistance must be even through the travel. "Dead" play indicates a faulty shock absorber, which must be replaced.

- Make sure that the shock absorber is the correct one for the vehicle in question, i.e., model year, carrying capacity, etc.

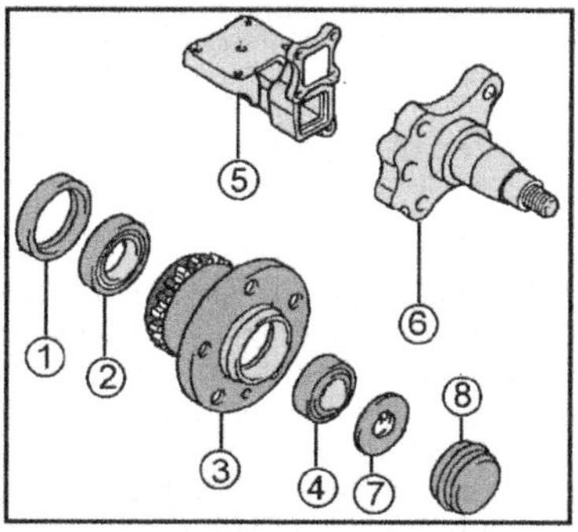

The installation is a reversal of the removal procedure, but the vehicle must be weighted down as described during the replacement of the rear springs, i.e., the dimension "H" in Fig. 7.5 must be set to the specific value. In this case, this can be obtained by placing weights in the cargo room of the vehicle, when the vehicle is resting on its wheels. Tighten both mountings to 16 kgm (115 ft.lb.).

Fig. 7.11 – Rear wheel hub and wheel bearing.
1 Oil seal
2 Inner wheel bearing
3 Wheel hub and toothed rotor (ABS)
4 Outer wheel bearing
5 Rear axle housing
6 Axle stump
7 Thrust washer
8 Grease cap

7.4. Rear Hubs and Wheel Bearings

The rear wheel bearings must be adjusted after the brake drums have been fitted. After the adjustment, the nut is peened in position, preventing it from coming loose. Fig. 7.11 shows the component parts of a wheel hub/wheel bearing assembly. The wheel hub shown is for a vehicle with ABS (toothed rotor fitted to the wheel hub).

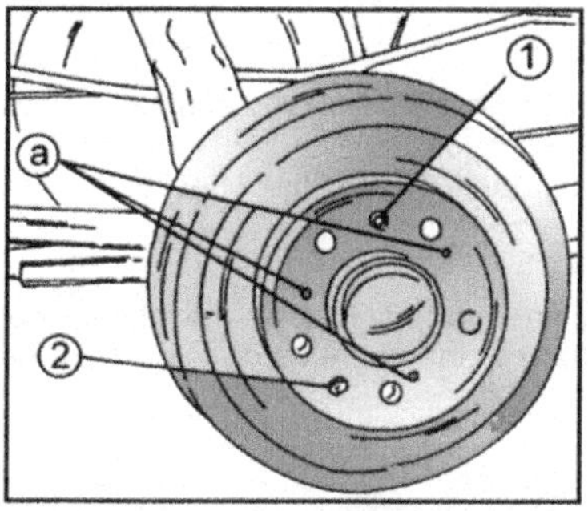

A wheel hub is removed as follows.

Fig. 7.12 – When removing a brake drum, unscrew the guide pin (1) and the screw (2). Screw three screws into the threaded holes "a" to remove a tight drum.

- Slacken the wheel bolts, place the rear end of the vehicle on chassis stands and unscrew the wheel or wheels.
- Unscrew the guide pin for the wheel (1) and the screw (2) in Fig. 7.12 and remove the brake drum. The brake drum will be very tight and you will need three M20 × 1.25 bolts to withdraw it. Screw the bolts into the threaded bores "a" and tighten them alternatively until the drum can be taken off.
- Remove the hub grease cap with a small chisel. We must warn you that the cap may be damaged.
- Find the peening of the wheel bearing nut (2 places) and knock it back. The wheel bearing nut must now be slackened and removed. Remove the thrust washer underneath the nut. Fig. 7.13 shows the operation at this stage.

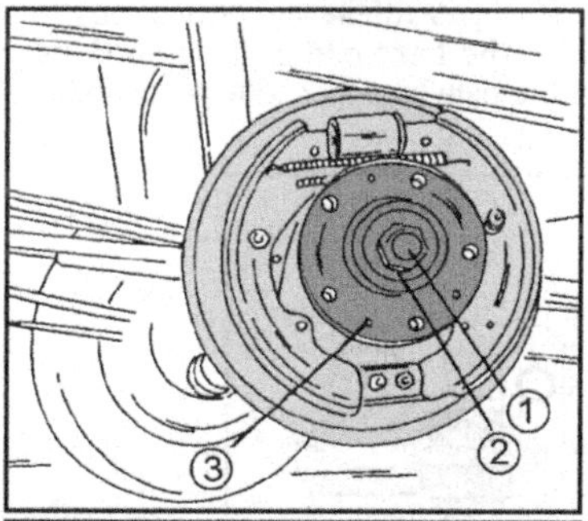

Fig. 7.13 – View of the fitted wheel hub.
1 Wheel bearing nut
2 Thrust washer
3 Wheel hub

- If the axle stump must be removed, remove the bolts (2) in Fig. 7.14. The axle stump (1) can now be removed.
- From the inside of the wheel hub, remove the wheel bearing. The outer bearing race is driven out of the wheel hub, if the bearing is to be replaced.
- Thoroughly clean all bearing parts. If the bearing requires replacement, dismantle the wheel hub as described above.

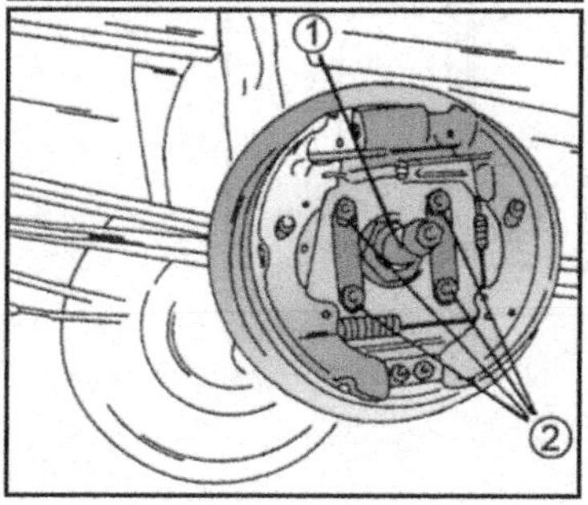

Assemble and refit a wheel hub as follows.

- Fill the inside of the wheel hub with heat-resisting grease (wheel bearing grease).
- Drive the two outer wheel bearing races from opposite sides into the wheel hub. Use a good drift to avoid damage.

Fig. 7.14 – View of the rear axle without brake drum. The axle stump (1) is secured with bolts (2).

- Insert the inner wheel bearing cage and fit a new oil seal.
- If the axle stump has been removed, refit it with new bolts. The bolts are tightened to 15 kgm (108 ft.lb.).
- Slide the wheel hub over the axle stump and drive it in position, using a soft-metal hammer or plastic mallet. Grease the outer bearing cage, insert it into the wheel hub and place the washer in position. Fit a new wheel bearing nut and tighten it finger-tight.

The wheel bearing must now be adjusted. Important is the bearing clearance of 0.025–0.10 mm. Proceed as follows.

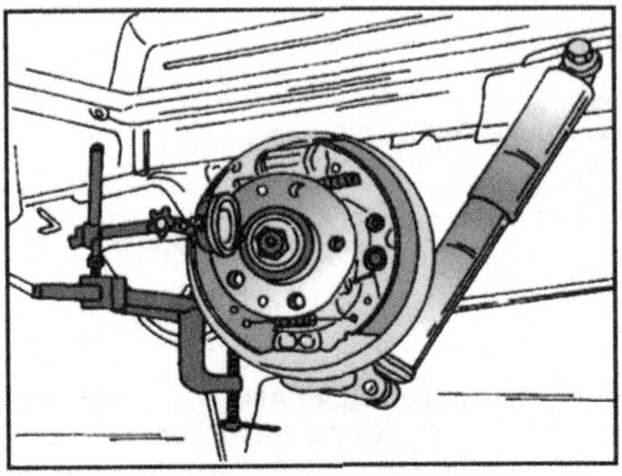

Fig. 7.15 – Checking the bearing clearance.

- Tighten the wheel bearing nut to 2.0 kgm (14.5 ft.lb.). Rotate the brake drum during the tightening process to settle the wheel bearings in their races.
- Completely slacken the nut and then re-tighten it again to 2.0 kgm.
- Fit a dial gauge as shown in Fig. 7.15, with the stylus resting on the outside of the wheel hub flange. Push the wheel hub fully in and set the dial gauge to "Zero". Pull the hub outwards and read off the indication on the gauge. The result is the end float of the wheel hub.
- If the clearance is not within the values given above, slacken or tighten the nut.
- In the final position, peen the wheel bearing nut into the axle stump to secure it. Do not split the metal of the nut collar.
- Remove the dial gauge.
- Fit the brake drum and tighten it in position by means of the guide pin and the two securing screws (see Fig. 7.12).
- Fit the hub grease cap and the wheel and lower the vehicle to the ground. Tighten the wheel bolts.

7.5. Rear Drive Shafts (4WD)

The rear drive shafts have a tri-axe joint on the axle side and a constant velocity joint on the wheel side. Both shafts are of the same length. The arrangement of the tow shafts is the same as described for the front wheel drive shafts.

7.6. Rear Axle (With 4WD)

The rear axle of a vehicle with four-wheel drive is secured by means of two rubber-metal bearings to a bracket on the body and two rubber-metal bearings to the sub-frame. The front end of the axle drive is connected to the propeller shaft. Fig. 7.16 shows a sectional view of the axle together with the location of the mountings.

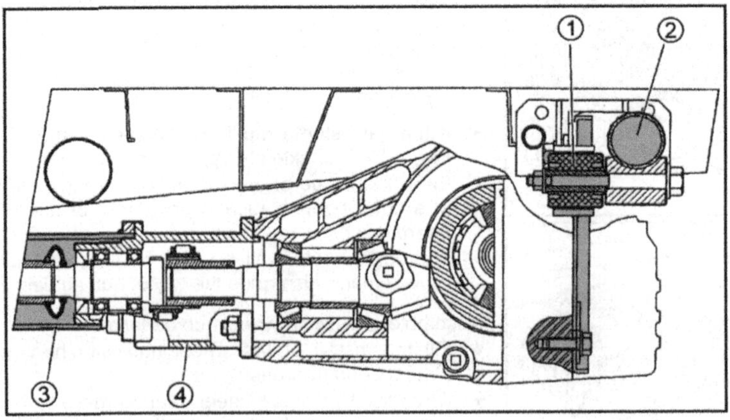

Fig. 7.16 – Sectional view of the rear axle (4 × 4).

1 Rubber/metal bearing on sub-frame
2 Sub-frame
3 Propeller shaft tube
4 Selector housing for 4 × 4

The oil level in the rear axle case should be checked every 6000 miles. Change the oil every 20 000 miles. The axle oil capacity is 2.3 litres (approx. 4 pints). An oil filler/level plug is inserted into the rear of the rear axle housing. A drain plug is fitted to the bottom of the housing. Warm oil will speed up the draining operation.

Make sure that the sealing washers of the plugs are in good condition before they are fitted.

Further information is given in Chapter 3 in conjunction with the transmission.

7.7. Propeller Shaft (4WD)

The propeller shaft consists of two parts. The front part is fitted with a universal joint and a sliding joint, which is inserted into the rear of the transmission to compensate for to and fro movements of the power unit. The rear propeller shaft consists of a shaft that rotates inside a tube.

Depending on the version, one or two intermediate bearings are fitted (vehicles with short or long wheelbase). The drive is transmitted to the propeller shaft when the four-wheel drive is engaged.

Fig. 7.17 shows the details of the propeller shaft mounting. Removal and installation can be carried out by referring to the illustration. It will be noted that the universal joints of the shaft and the sliding joint have grease nipples (depending on the length of the shaft one or two nipples in the joints). The nipples should be filled every 6000 miles with a suitable grease (grease gun).

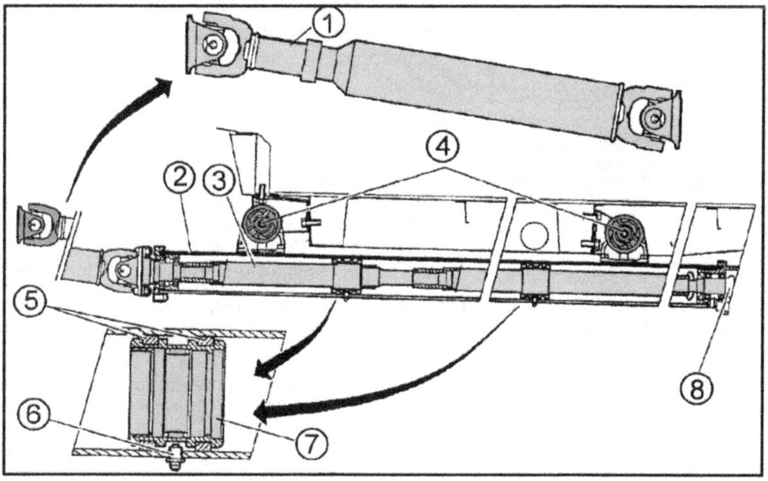

Fig. 7.17 – Arrangement of the propeller shaft (4 × 4). The arrows show the location of the grease nipple.

1 Propeller shaft
2 Propeller shaft tube
3 Propeller shaft with 3 intermediate bearings (long wheelbase)
4 Intermediate bearing (long wheelbase)
5 Rubber bearing mounting
6 Grease nipple
7 Mounting bearing
8 Splined end

7.8. Rear Axle Tightening Torques

Shock absorber mountings .... 16.0 kgm (115 ft.lb.)
Rear spring mounting, front and rear .... 15.5 kgm (112 ft.lb.)
Spring "U" bolt nuts .... 13.0 kgm (94 ft.lb.)
Spring hanger brackets to chassis .... 14.0 kgm (101 ft.lb.)
Spring shackle mounting to spring and chassis .... 15.5 kgm (112 ft.lb.)
Rear axle shaft stump to axle .... 15.0 kgm (108 ft.lb.)
Rear hub nut .... 2.0 kgm, then adjust and lock

Wheel Bolts:
- Models 10Q and 14Q .... 16.0 kgm (115 ft.lb.)
- Model 18Q .... 18.0 kgm (120 ft.lb.)

7.9. Coil Spring Rear Suspension

7.9.1. REAR AXLE – REMOVAL AND INSTALLATION

The crossmember for the rear suspension must be guided in position with a guide bolt during installation. Make up such a belt as described below.

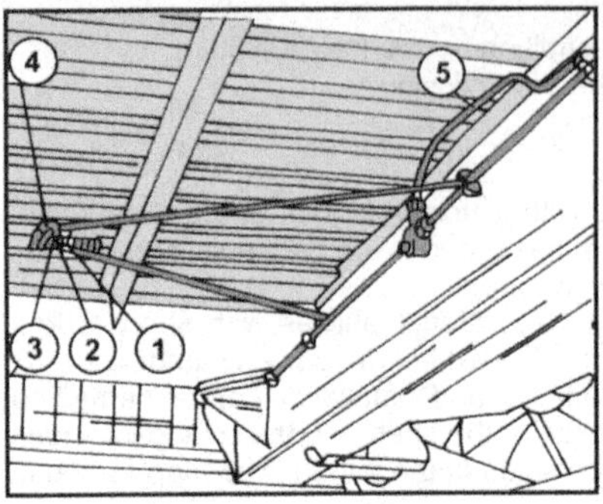

Fig. 7.18 – Details for the removal and installation of the handbrake cable and the brake hose (refer to the text).

- Place the rear end of the vehicle onto chassis stands. The wheels remain in position as the axle has to be "rolled" out during removal. The wheels must be suspended under their own weight.
- The following operations are carried out by referring to Fig. 7.18 when disconnecting the handbrake cable. Slacken the locknut (1) and the nut (2) on the handbrake compensating lever, remove the inserted sleeve (3) and withdraw the compensating lever (4) from the handbrake cable. On the other side, disconnect the brake hose (5) from its connection on the pipe. Free the pipe from the clip on the crossmember. Plug the open connections to prevent entry of dirt.

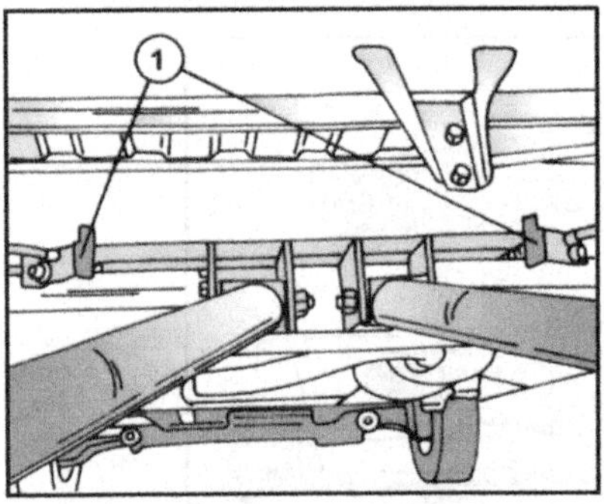

Fig. 7.19 – Remove the two mounting brackets (1) on the L.H. and R.H. sides at the positions shown.

- Refer to Fig. 7.19 and remove the two mounting brackets (1) on the L.H. side and on the R.H. side.
- Disconnect the linkage (5) in Fig. 7.3 at the upper and lower ends. The lower end is secured by means of a spring pin, which must be replaced during installation.

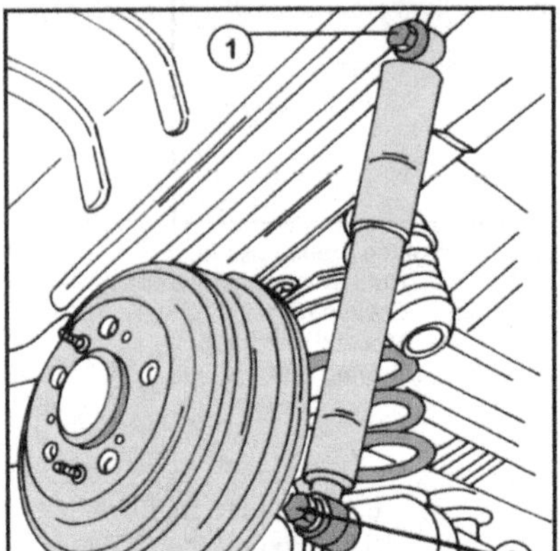

Fig. 7.20 – View of a fitted shock absorber. The bolts (1) and (2) are tightened when the dimension "H" in Fig. 7.24 is obtained. A spacer sleeve is fitted to the inside of the lower damper mounting.

- Place a mobile jack underneath one of the suspension arms until the shock absorber is compressed, i.e., the suspension arm will be approx. in its normal position. Remove the upper bolt (1) in Fig. 7.20. The lower bolt can be left on the suspension arm or is removed in order to take out the damper. In this case, remove the spacer sleeve on the inside of the shock absorber mounting.
- Slowly lower the jack and remove the coil spring.
- Carry out the same operations on the other side of the vehicle.
- The rear suspension crossmember can now be removed. Check that the rear suspension is securely supported on the jack, and on both sides of the crossmember, remove the bolts (1) in Fig. 7.21.
- Lower the complete rear suspension onto the rear wheels and "roll" it from under the vehicle. Remove the rubber mountings.

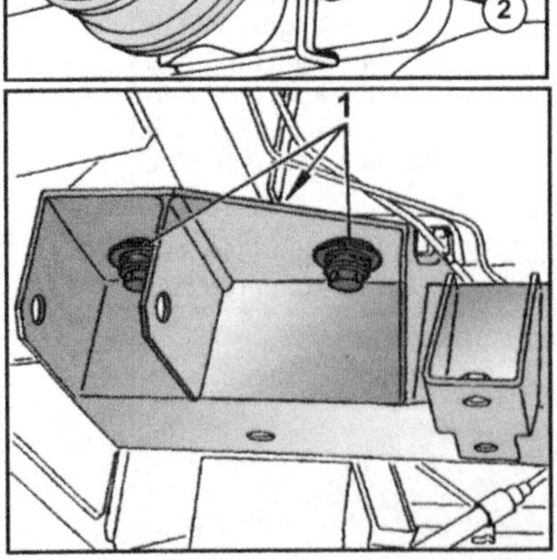

Fig. 7.21 – The bolts (1) secure the crossmember to the body. They must be tightened when the weight of the vehicle is on the rear wheels.

The installation is carried out as follows. Remember that you will need the guide bolts already mentioned.

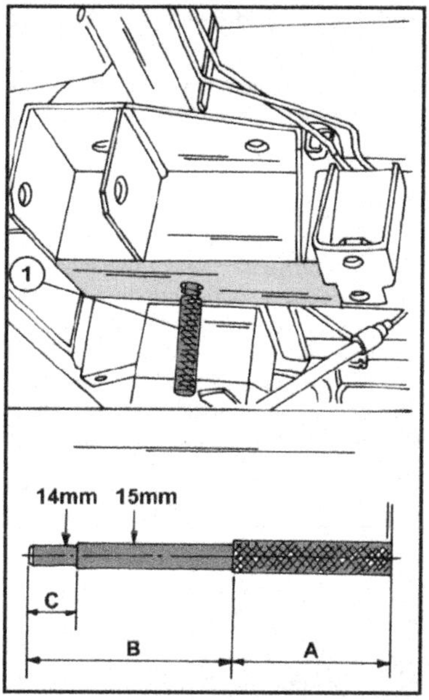

Fig. 7.22 – A guide bolt (1) is inserted on each side to centre the crossmember on the body. The lower view shows the dimensions of the guide bolts, which must be made-up accordingly.
A = 100 mm
B = 130 mm
C = 30 mm

- Fit the rubber bearings and lift the crossmember into its approximate position until it nearly touches the body. Now insert one of the guide bolts made-up to the dimensions shown in Fig. 7.22 into the crossmember and the body.
- Fit the securing bolts (1) in Fig. 7.21 and then tighten finger-tight.
- Place a jack underneath the suspension arm and fit the coil spring onto the arm. Lift the suspension arm until the spring is in its approximate position. Check that the upper end lower ends of the spring are engaging properly. Fit the shock absorber to the upper end (or refit it completely if it has been removed), without tightening any of the bolts. Fit the second spring in the same manner. The bolts are tightened as described below.
- Refit the brake pressure regulator linkage, remembering to replace the spring pin at the bottom.

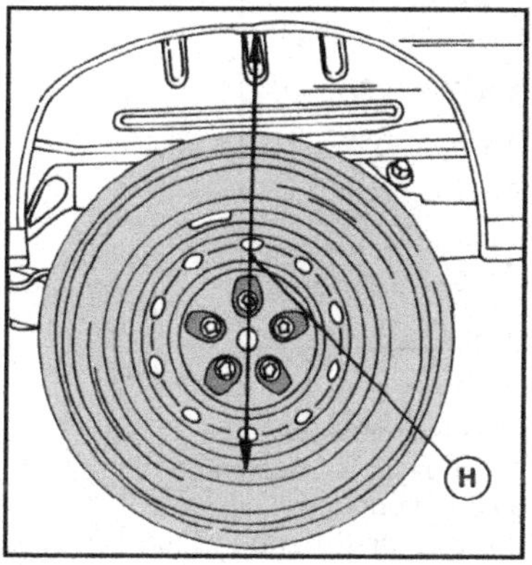

Fig. 7.23 – Before tightening the shock absorber, securing bolts load down the rear end of the vehicle to obtain dimension "H" as given in the text.

- Re-connect the handbrake cables and the brake hose and attach the brake pipe with the clip to the crossmember.
- Lower the vehicle onto its wheels and tighten the bolts (1) in Fig. 7.21 to 12.0–13.0 kgm (86–94 ft.lb.). The two guide pins can now be removed.
- The rear end of the vehicle must now be loaded down to pre-tension the shock absorbers. Dimension "H" in Fig. 7.23 must be exactly 620 mm. In this position of the rear suspension, tighten the shock absorber bolts to 17.0 kgm (122 ft.lb.). A tolerance of 10% more or less is permissible.
- Bleed the brake system and adjust the handbrake cables as described in the "Brakes" section.

7.9.2. SHOCK ABSORBERS – REMOVAL AND INSTALLATION

- Place the rear end of the vehicle on chassis stands. The wheels must hang down under their own weight.
- Place a mobile jack or a hydraulic jack underneath the suspension arm and lift up the arm until the shock absorber is compressed. Remove the bolts (1) and (2) in Fig. 7.20 and lift out the shock absorber. At the inside of the lower mounting, remove the spacer sleeve, now free.

The installation is a reversal of the removal procedure, but remember that the rear end of the vehicle must be loaded down in order to tighten the shock absorber mounting bolts, i.e., the dimension "H" in Fig. 7.23 must be established before the bolts are tightened.

7.9.3. SUSPENSION ARMS – REMOVAL AND INSTALLATION

- Slacken the wheel bolts, jack up the rear end of the vehicle and remove the wheel.
- Remove the brake drum as described in the "Brakes" section.
- Disconnect the handbrake cable by referring to Fig. 7.24 (see also Fig. 7.18). Using a suitable drift, drive out the small guide sleeve from the inside of the brake back plate and withdraw the handbrake cable from the rear of the plate.

- Disconnect the brake pressure regulator linkage (5) in Fig. 7.3 at the lower end (always replace the spring pin) and push it to one side. This operation is carried out on the R.H. side of the vehicle.

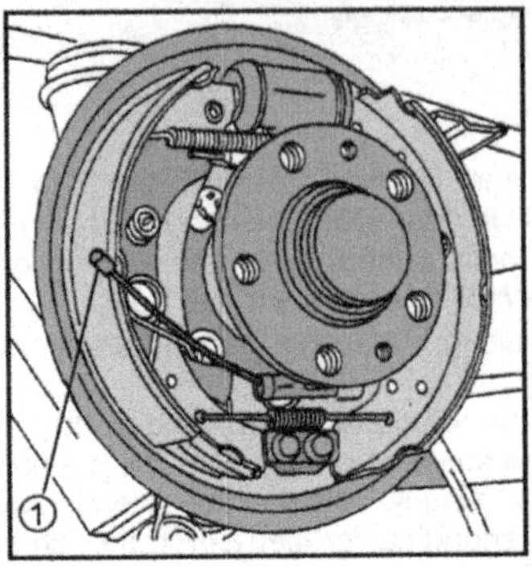

Fig. 7.24 – Removal of the handbrake cable (1) from the inside of the rear brake. The guide sleeve is driven out of the brake back plate.

- Remove the shock absorber and the coil spring as already described.
- The remaining operations are carried out by referring to Fig. 7.25. Slacken the unit nut securing the brake pipe (5), drive out the spring retaining plate (4) and disconnect the pipe from the brake hose. Close off the open connections. On the suspension arm, remove the bolt (2) and the lock (3). On the other side, remove the nut (7) and take off the washer (6). Drive out the bearing bolt, using a suitable drift. The bolt has a flat at the lower end (marked with "a"), which must be in the same position when the arm is refitted. The suspension arm can now be removed.

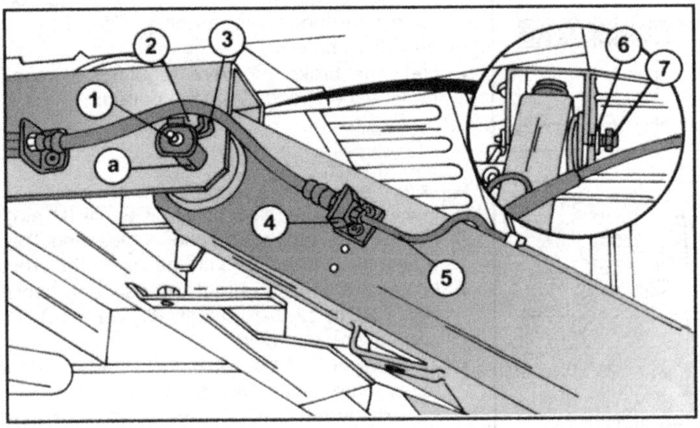

Fig. 7.25 – Details for the removal and installation of a rear suspension arm. The numbers are referred to the text.

The installation is a reversal of the removal procedure. The following points should be noted.

- Coat the bearing pin (1) in Fig. 7.25 with grease (total "G6" grease is recommended by the manufacturer). The nut is tightened when the wheels are resting on the ground.
- Fit the suspension arm and attach it with the bearing pin (1), the washer (6) and the nut (7). Only tighten finger-tight.

Note: Take care that the flat "a" is in the correct position, i.e., at the bottom.

- Connect the brake hose (8) to the brake pipe and attach the assembly with the spring plate.
- Refit the coil spring and the shock absorber as described above.
- Re-connect the linkage (5) in Fig. 7.3 (if removed, R.H. side only). Remember to replace the spring pin.
- Refit the coil spring and the shock absorber as described above.
- Re-fit the handbrake cable and connect it to the handbrake operating lever, and refit the brake drum and the wheel.
- Lower the vehicle to the ground and tighten the wheel bolts. The nut (7) is now tightened to 11.0–13.0 kgm (79–94 ft.lb.).

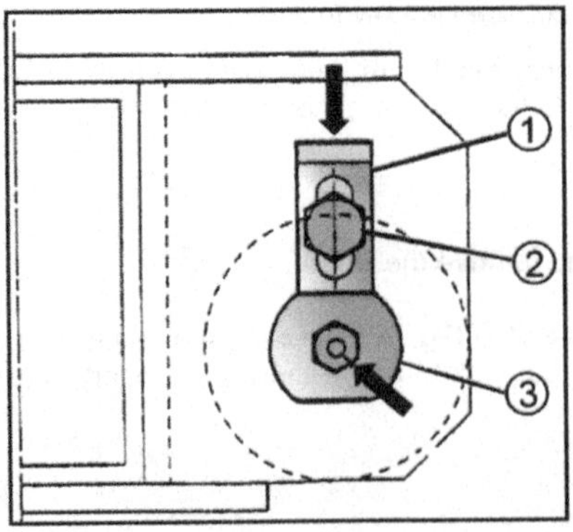

Fig. 7.26 – The bearing pin for the suspension arm must be locked against rotation as described in the text.

The bearing pin must now be prevented from rotation. This is carried out by referring to Fig. 7.26.

- Tighten the lock (1) finger-tight with the bolt (2) and knock downwards until it contacts the bearing pin (3).
- In this position, tighten the bolt (2) to 4.0 kgm (30 ft.lb.).

Finally, bleed the brake system as described in the "Brakes" section.

8. BRAKE SYSTEM

8.0. Technical Data

Type fitted	Dual-circuit system. System without ABS diagonally split, system with ABS split front/rear circuit. With load-sensitive brake pressure regulator, brake servo unit. Handbrake acting on rear wheels. ABS fitted as optional extra.
Front Brakes	
Type	Lucas (Girling) sliding-type brake calipers with two pistons of different diameters. Calipers different for the various models.
Brake discs:	
- Models 10Q and 14Q	Solid discs
- Model 18Q	Ventilated discs
Brake Cylinder Diameter:	
- Models 10Q and 14Q (10 and 14)	40 and 48 mm (1.59 and 1.9 in.)
- Model 18Q	45 and 48 mm (1.79 and 1.9 in.)
Brake Disc Diameter:	
- Models 10Q and 14Q (10 and 14)	280.0 mm (11.12 in.)
- Model 18Q (Maxi)	300.0 mm (11.92 in.)
Brake Disc Thickness:	
- Models 10Q and 14Q (10 and 14)	18.0 mm (0.72 in.)
- Model 18Q (Maxi)	24.0 mm (0.95 in.)
Min. permissible thickness:	
- Models 10Q and 14Q	15.9 mm (0.63 in.)
- Model 18Q	21.9 mm (0.87 in.)
Brake pad thickness (all models)	18.2 mm (0.72 in.)
Min. brake pad thickness (without metal plate)	1.0 mm (0.04 in.)
Brake disc run-out	0.15 mm (0.006 in.), measured 2 mm (0.08 in) from the outer edge.
Rear Drum Brakes	
Type	With leading and trailing brake shoes, automatic adjustment
Wheel Brake Cylinder Diameter:	
- Models 10Q and 14Q, without turbo	25.0 mm (0.99 in.)
- Models 10Q and 14Q, with turbo	27.0 mm (1.07 in.)
- Model 18Q	28.6 mm (1.14 in.)
Brake drum diameter, all models	254.0 mm (10.09 in.)
Max. brake drum diameter, all models	255.6 mm (10.15 in.)
Min. thickness of brake linings	1.0 mm (0.04 in.)
Master Brake Cylinder:	
- Piston diameter – models 10Q and 14Q	22.2 mm (0.88 in.)
- Piston diameter – model 18Q	25.4 mm (1.0 in.)
- from chassis number 15 091 011	23.81 mm (0.95 in.) – only 10Q/14Q
Brake Fluid:	
- Type	SAE J1703 Dot 4
- Regular replacement	Every 25 000 miles or 2 years

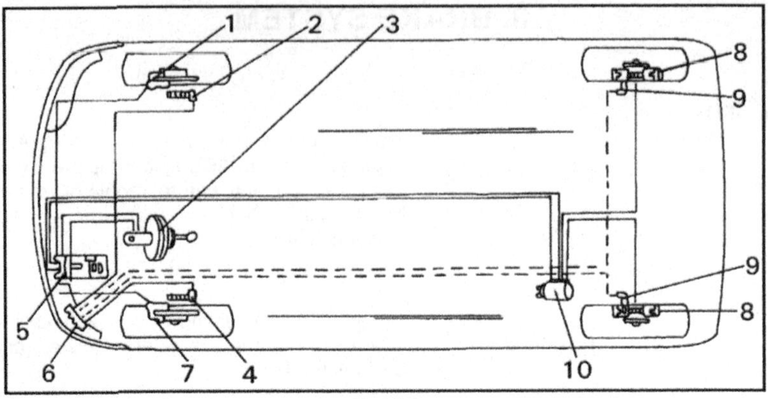

Fig. 8.1 – Layout of the brake system component parts with ABS.

1 R.H. front brake caliper
2 R.H. front wheel speed sensor
3 Master brake cylinder with brake servo
4 L.H. front wheel speed sensor
5 ABS hydraulic control unit
6 Electronic ABS control unit
7 L.H. front brake caliper
8 Brake drums
9 Rear wheel speed sensor
10 Brake pressure regulator

8.1. Short Description

All models covered in this manual have disc brakes at the front. The brakes are made by Girling. Each front disc brake assembly has two pistons of different diameters, supplied independently by separate circuits. Sliding calipers are employed, i.e., the pressure is applied to the pistons on one side and the complete brake caliper is moved sideways on slide bolts to press the brake pad on the opposite side against the brake disc. Different calipers are fitted to models 10Q/14Q and 18Q. The smaller of the two pistons has a larger diameter on the 18Q model. Also, note different disc diameters and disc thicknesses.

The tandem master cylinder works on the dual-line principle, but there are differences. On models without ABS, the system is diagonally split. The rear circuit of the cylinder supplies the upper piston of each front caliper and the rear brakes. The front circuit of the cylinder supplies the lower pistons of each caliper. The front and rear circuits are split if ABS is fitted, i.e., one circuit supplies the front brakes and the other one supplies the rear brakes. Fig. 8.1 shows the layout of a circuit with ABS. The circuit without ABS looks similar, with the difference that the parts belonging to the ABS system are not shown.

Not all models have the same master cylinder, differing in the diameter. If a cylinder must be changed, make sure that the correct part is fitted.

A brake servo unit is fitted, the vacuum being supplied by a separate vacuum pump, driven by the engine camshaft. A brake pressure regulator regulates the brake pressure to the rear wheels. Fig. 8.2 shows the location of parts.

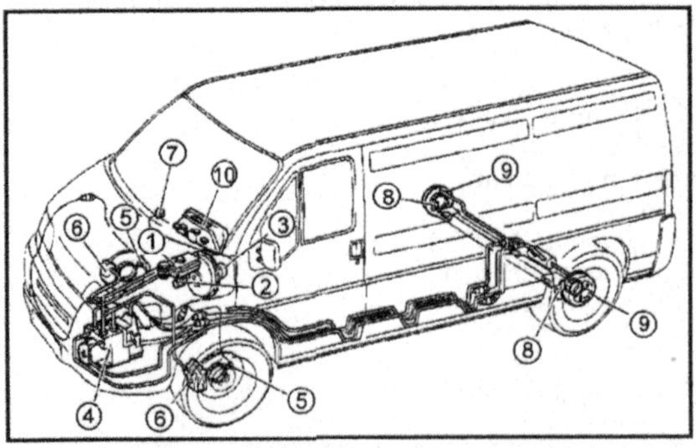

Fig. 8.2 – The location of the various component parts of a Bosch ABS system.

1 Fluid reservoir
2 Brake master cylinder
3 Brake servo unit
4 ABS control unit
5 Front wheel speed sensor
6 Front brake caliper
7 ABS warning light
8 Rear wheel speed sensor
9 Rear brake drum
10 Brake warning light

The handbrake operates the rear wheels. The rear wheels should be locked when the handbrake lever is pulled 4–5 notches.

8.2. Front Disc Brakes

8.2.0. REPLACING THE BRAKE PADS

To check the remaining thickness of the brake pad material, remove the front wheels and check the two brake pads should the inspection "window". Always check on both sides, as uneven wear is possible. If this is the case, check the brake caliper on the side with the increased wear thoroughly, as the caliper could be sticking. If the thickness is less than 4 mm, replace the brake pads.

If the brake pads are removed for other reasons than replacement, mark each pad as you take it out, i.e., L.H. and R.H. sides and inner and outer pads. Incorrect fitting could lead to brake pull to one side.

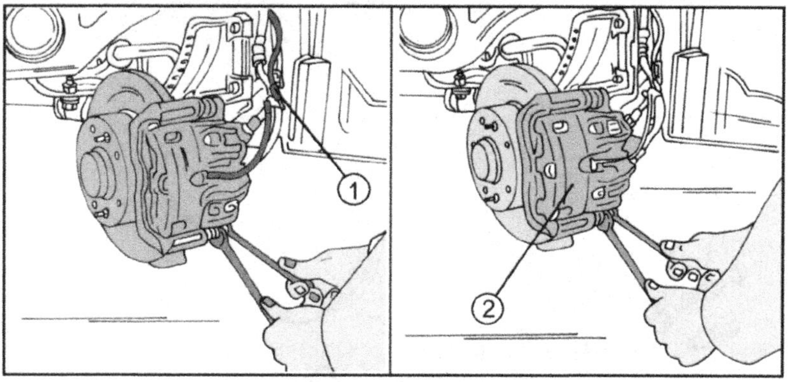

Fig. 8.3 – Removal of a brake caliper (2). When removing the slide bolts, counterhold the hexagon at the inside with an open-ended spanner. It may be possible to ground down the spanner to insert it as shown. The pad wear indicator cable is connected at position (1).

Replace the brake pads as follows.

- Slacken the wheel bolts on both sides, jack up the front end of the vehicle and place chassis stands underneath the sides of the body. The suspension must be allowed to hand down under its own weight.
- Disconnect the cables from the pad wear indicator on the side of the spring strut/brake caliper (Fig. 8.3, L.H. view).
- Grip the brake caliper and with a sharp pull, move it towards the outside. This will force the piston back into their bores. Care must be taken when the fluid reservoir is fairly full, as it can overflow. Check and, if necessary, draw off some fluid.
- Using a ring spanner and an open-ended spanner as shown in Fig. 8.3 on the R.H. side and unscrew the slide bolts. The lower bolt is removed; the upper bolt is slackened. The two bolts are self-locking and must be replaced. In emergency, it is, however, possible to use "Loctite" during installation.
- Swing the caliper towards the top, as shown in Fig. 8.4, and remove the inner and outer brake pads.

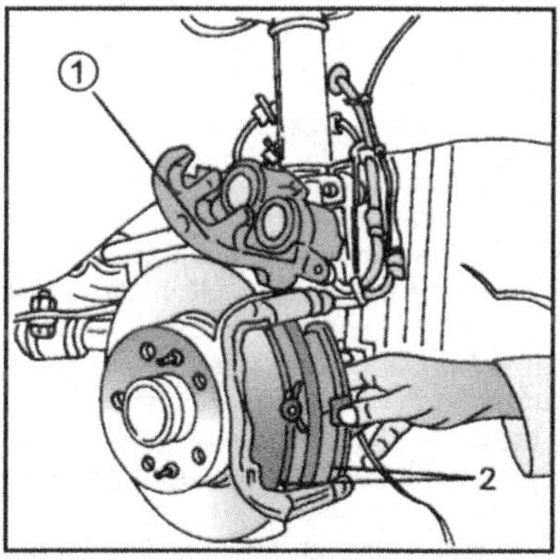

Fig. 8.4 – Removal of a brake caliper. The caliper (1) is moved upwards and tied up. The brake pads (2) can be removed from both sides of the brake disc.

Fig. 8.6 shows a detailed view of the removed brake caliper together with the brake pads.

Important: The brake pedal must not be depressed after the brake pads have been removed.

Clean the inside of the brake caliper opening with a stiff brush to remove brake dust. The pistons must now be pressed into their bores, if they appear to be protruding above the caliper bore. Use a hammer handle but have a helper observing the fluid level in the reservoir, as the fluid will rise, as the pistons are pushed in. If necessary, remove some fluid out of the reservoir or open the bleed screw whilst the first piston is pushed in. Careful handling will not allow air into the system – otherwise, the brake system must be bled of air.

If the sealing rubbers for the slide bolts are in bad condition, unscrew the guide bolt. Clean the bolt and refit it with the new rubber part. The guide bolt is fitted with rubber grease.

The installation of the brake pads follows the description below.

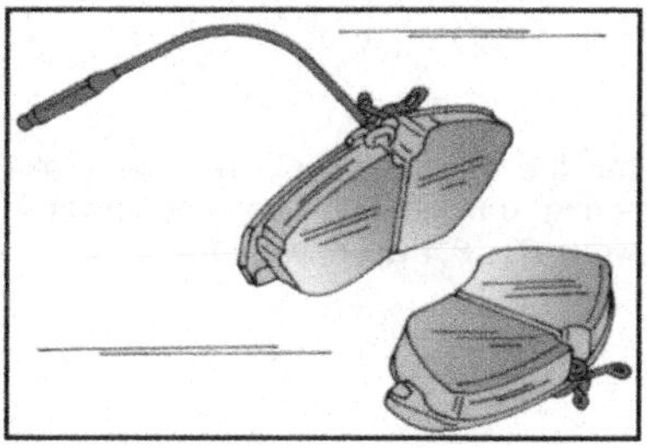

- Insert the inner brake pad (with the cable for the pad wear indicator), making sure that the cable is properly routed. Now fit the outer brake pad.

Fig. 8.5 – View of the two brake pads. The pad with the pad wear indicator cable is fitted to the inside.

- Check once more that the brake pads are fitted correctly and swing the caliper over the brake pads. Fit the new bolt at the bottom of the caliper. New bolts are normally included in the brake pad repair kit. *If none is supplied, coat the threads with "Loctite".*

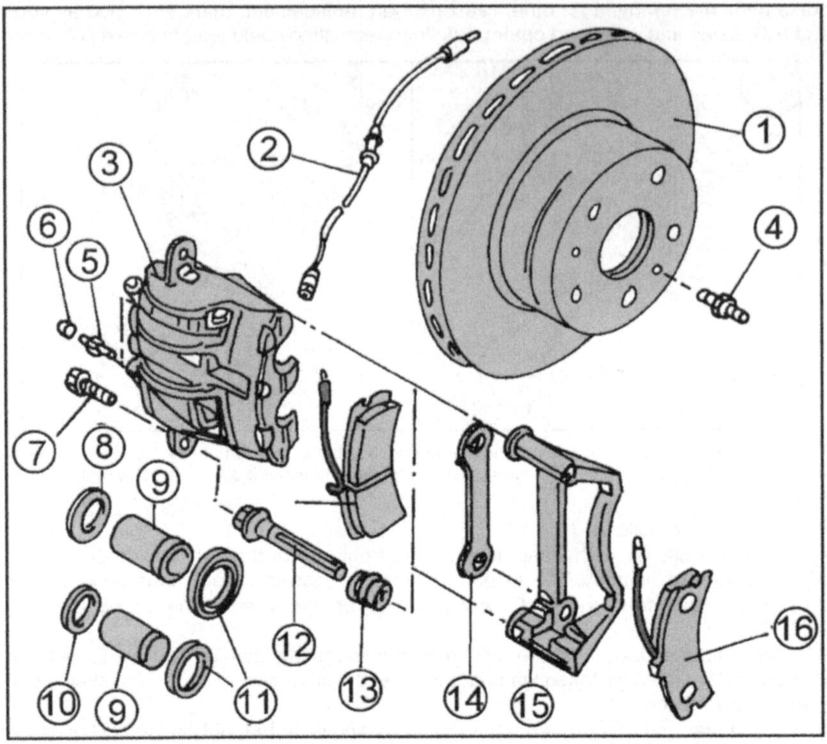

Fig. 8.6 – Exploded view of a brake caliper with two pistons (without ABS).

1 Ventilated disc
2 Brake pad wear indicator cable
3 Brake caliper cylinder
4 Brake disc screw
5 Bleed screw
6 Rubber dust cap
7 Guide bolt, 2.7 kgm
8 Cylinder seal, upper piston
9 Brake cylinder piston
10 Cylinder seal, lower piston
11 Rubber dust seal
12 Slide bolt
13 Rubber grommet
14 Plate
15 Caliper mounting bracket
16 Brake pad set

- Remove the upper slide bolt, coat the threads with "Loctite" and refit it into the caliper.
- Now tighten both bolts to 2.7 kgm (20 ft.lb.). First, tighten the lower bolt. Again, hold the hexagon at the inside with an open-ended spanner, as shown in Fig. 8.3. Now tighten the upper bolt in the same manner.
- Re-connect the cable for the brake pad wear indicator (Fig. 8.3, left). Check that the cables cannot touch other parts of the front suspension, when the wheels are turned from lock to lock.
- Operate the foot brake a few times to set the new pads against the brake discs. You will also find out, if the brake system requires bleeding. Pump the pedal with quick strokes. A rising pedal normally indicates air in the system.
- Refit the wheels and lower the vehicle to the ground. Tighten the wheel bolts with the specified torque (depending on the vehicle models).
- Check and, if necessary, correct the fluid level in the reservoir.

Remember that new brake pads need a while to "bed" in, before the brakes obtain the full performance. Treat the brakes with care when you first drive off.

8.2.1. BRAKE CALIPER – REMOVAL AND INSTALLATION

The removal and installation of a brake caliper follows, in general, the description above. The brake caliper cylinder can then be unscrewed from the suspension steering knuckle (the two outer bolts in Fig. 8.7) or is free after both slide bolts are removed. The parts shown in Fig. 8.4 will now be free.

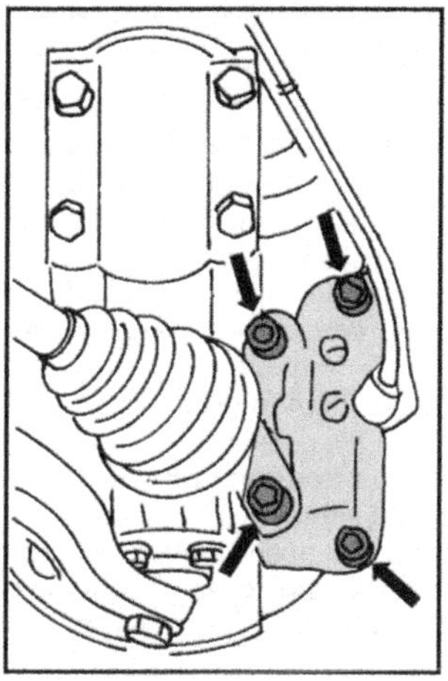

Fig. 8.7 – Attachment of a brake caliper on the steering knuckle. The bolts on the right show the location of the slide bolts, and the bolts on the left are the bolts for the mounting bracket.

As two brake hoses are fitted, separate the connection between the hoses and the pipes, i.e., it is not possible to remove the brake caliper and then the brake hose. Brake pipes/brake hoses are connected with union nuts and spring plates. Close the ends of the brake pipes in a suitable manner to prevent entry of dirt.

If the brake caliper mounting bracket is to be removed, unscrew the two inner bolts in Fig. 8.7. Below the mounting frame is a washer, which must not be omitted during installation.

Note the following during installation, also assuming that the mounting bracket has been removed.

- Fit the brake hoses to the caliper body and tighten them to 1.3 kgm (11 ft.lb.).
- Thoroughly clean the mounting face of the caliper and steering knuckle, coat the caliper bolts with "Loctite", fit the mounting bracket in position and fit the bolts. Tighten the bolts in several stages to a torque of 21 kgm (151 ft.lb.).
- Fit the brake pads as already described.
- Insert the brake hoses into the bracket and secure with the spring plates. Fit the brake pipes and tighten the union nuts. Move the wheels from one lock to the other and check that the brake hoses cannot rub against other parts of the front suspension. If necessary, slacken the union nuts and rotate the hose(s) accordingly.
- Finally, bleed the brake system as described later on. Operate the foot brake a few times before driving off.

8.2.2. BRAKE CALIPER OVERHAUL

Note the following general rules when overhauling brake calipers or for that matter any other hydraulic parts.

- The work must be carried out on a clean workbench.
- All rubber cups and seals must be replaced if the part has been dismantled, even if rubber parts still look useable, replace them. Before dismantling a caliper, make sure that repair kits are available.
- Never refit a piston or caliper if corrosion or other blemishes can be found. Always replace the part in question. Only use alcohol, white spirit or brake fluid to clean parts. Even a small amount of petrol can damage the rubber parts sooner or later. Any marks left by the use of white spirits can be cleaned-off with a lint-free cloth.

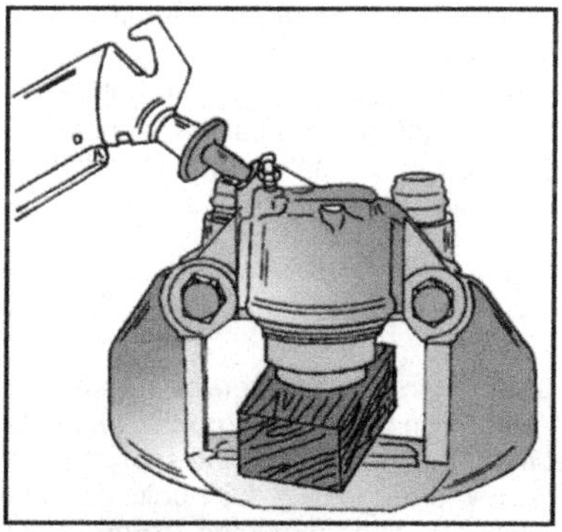

Fig. 8.8 – Removal of a piston. A wooden block is placed underneath the piston.

A caliper can be dismantled as follows.

- Unscrew the two brake hoses from the caliper and remove the two bleeder screws.
- Remove the dust seal(s) from the piston bores.
- The pistons are blown out. Place a piece of hardwood into the caliper opening to allow the piston to "hit" the wood and not the metal (see Fig. 8.8). Remember that you have two pistons, i.e., it may be necessary to deal with one piston at the time. Keep the fingers away from the area of the piston opening when the piston is ejected. You may be able to carry out the operation at a petrol station.

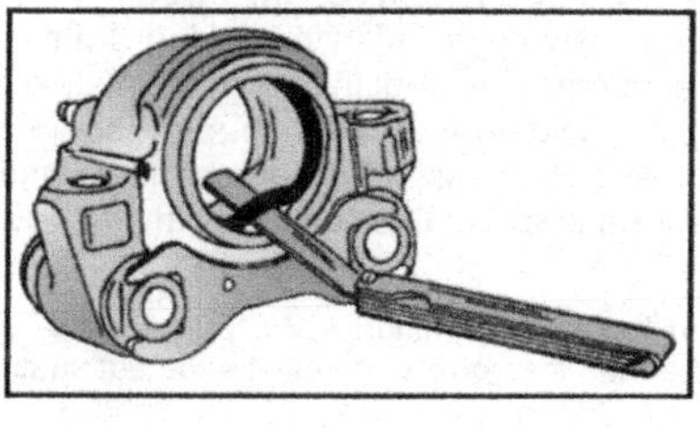

Fig. 8.9 – Removal of a cylinder sealing ring.

- From the inside of the cylinder bore, remove the cylinder sealing ring(s) with a blunt, pointed instruments, as, for example, shown in Fig. 8.9. Do not damage the bore(s).
- Clean all parts in clean brake fluid or alcohol. All parts in the repair kit must be used, i.e., do not use any of the old parts.
- Coat the cylinder seal(s) with clean brake fluid and insert them into the grooves(s), using the fingers only.
- Fit the dust seals to the pistons and slowly insert the piston into the caliper bore. Insert the seal lip into the cylinder grove and push the piston fully in. Check that the dust seals are correctly fitted.

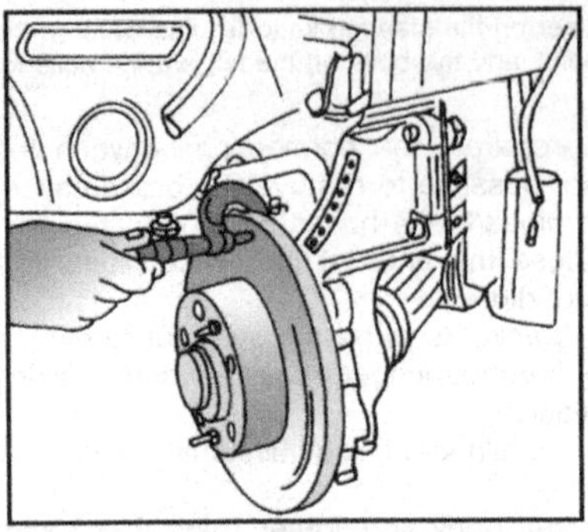

Fig. 8.10 – Measuring the thickness of a brake disc with a micrometre.

8.2.3. BRAKE DISCS

Brake discs must be replaced, if the friction faces show grooves or other damage. Grooves can develop if the brake pads have been allowed to wear down to the metal plates. Measure the thickness of the brake discs with a micrometre, as shown in Fig. 8.8 and compare the results with the dimensions given in Section 1.8.0. If in doubt, you may be able to get information for your particular vehicle from your dealer. If the discs show slight wear, you may have them re-ground.

Remove brake discs as follows.

- Slacken the wheel bolts, place the front end of the vehicle on chassis stands and remove the wheel.
- Remove the brake pads as described and remove the brake caliper completely. Also, remove the caliper mounting bracket. Tie the complete assembly with a piece of wire to the front suspension.

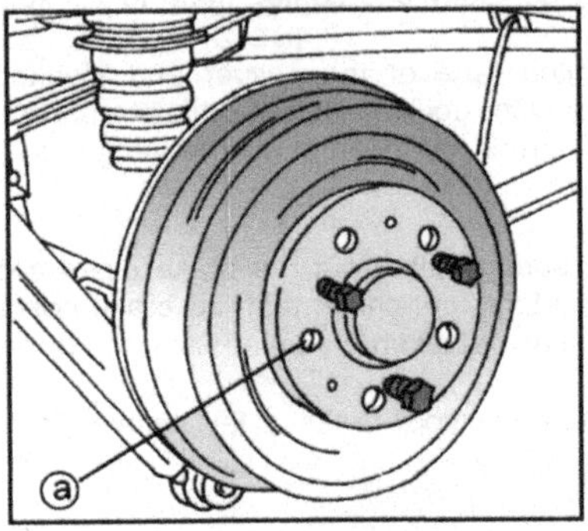

Fig. 8.11 – Brake discs are secured with guide studs at the positions shown. Rear brake drums are secured in a similar manner.

- Remove the two guide bolts from the outside of the wheel hub, as shown in Fig. 8.11. The brake disc has a tight fit on the hub. M10 × 1.25 bolts can be screwed into the threaded bores and used as extractor bolts to pull off the disc. Sometimes, it is enough to knock off the disc with a rubber mallet.
- During installation, fit the disc over the hub and tap it in position with a rubber or plastic mallet. Fit the guide bolts and tighten them alternatively to 1.0 kgm (7.2 ft.lb.).

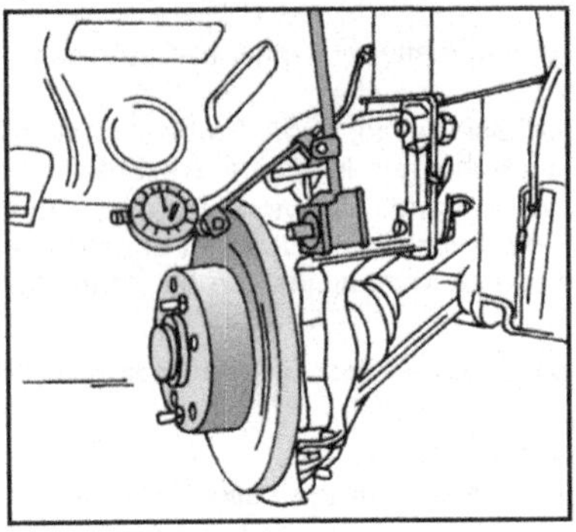

Fig. 8.12 – Checking a disc brake for run-out.

- Arrange a dial gauge with a suitable holder as shown in Fig. 8.12 and apply the stylus approx. 2 mm from the outer edge of the disc. Slowly rotate the wheel hub/disc and observe the dial gauge reading. The highest run-out is 0.15 mm (0.006 in.). Higher readings can be due to dirt between disc and hub. In this case, remove the disc once more and check the surfaces.
- Refit the brake caliper assembly, noting the tightening torque for the mounting bracket and the caliper slide bolts. Fit the wheel, lower the vehicle to the ground and tighten the wheel bolts.

8.3. Rear Brakes

8.3.0. BRAKE DRUMS – REMOVAL AND INSTALLATION

Normally, the brake drums can be removed after the removal of two guide bolts, the same as shown in Fig. 8.11 on the example of the brake discs. Sometimes, it is, however, necessary to pull off the drums. To do this, screw three M10 × 1.25 bolts into the holes shown in Fig. 8.13. By tightening the three bolts alternatively, the drum will slowly come off. If a sticking drum is experienced, release the handbrake and insert a screwdriver into the hole "a" in Fig. 8.11. The screwdriver will push against the adjuster lever of the automatic shoe adjustment and disengage a pin from the brake shoe. The brake drum will now come off as described.

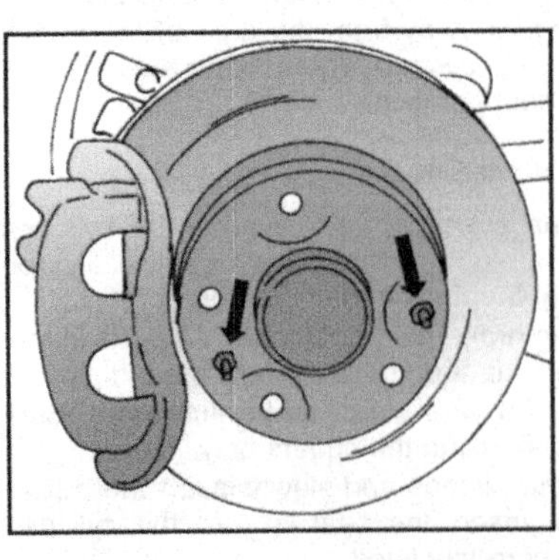

Fig. 8.13 – Three bolts can be used to withdraw the brake drum as described. Insert a screwdriver into the bore "a" to release the automatic adjustment lever.

8.3.1. REPLACING THE BRAKE SHOES

All models have the same rear brakes. After removal of the brake drum, make a sketch of the fitted shoe return springs to facilitate the installation. Specially note which way the spring hooks are inserted.

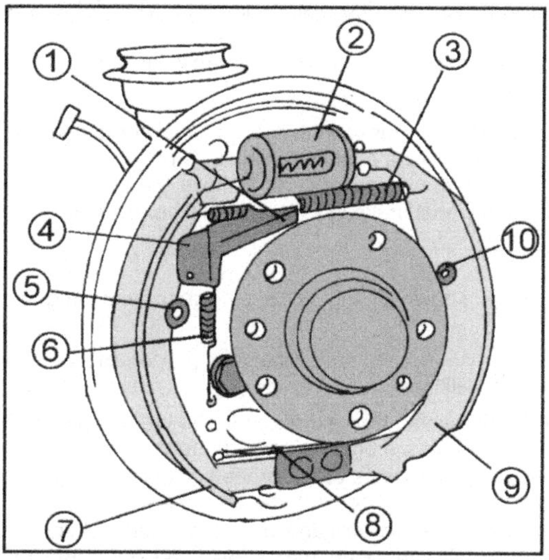

Fig. 8.16 – View of a rear brake assembly.
1 Adjuster strut
2 Wheel brake cylinder
3 Upper return spring
4 Adjuster lever
5 Brake shoe hold-down pin
6 Lever return spring
7 Brake shoe
8 Lower return spring
9 Brake shoe
10 Brake shoe hold-down pin

Brake shoes must only be replaced in sets. Never fit brake shoes of different manufacturers or with different lining materials. When we are talking brakes, we are talking original spare parts from the manufacturer. Fig. 8.14 shows a view of a brake assembly after the drum has been removed. Fig. 8.15 shows an exploded view of the brake assembly, which will help during removal and installation of the brake shoes.

- Disengage the return spring (6) with a pair of pliers and remove the adjuster mechanism (4).
- The upper return spring (3) and the lower return spring (8) must be removed with a pair of pliers or use a screwdriver and carefully lift the spring hooks out of their anchorage. The thrust piece, i.e., the strut for the self-adjusting mechanism of the brake shoes, in the centre can be released and can be taken out.

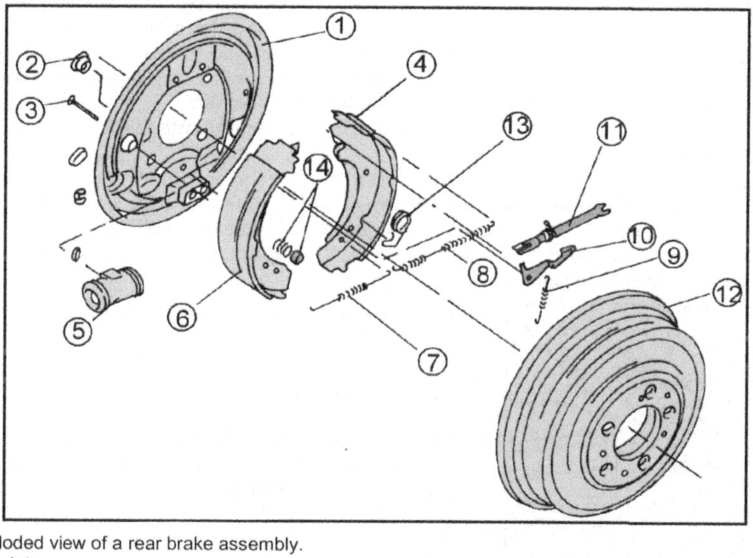

Fig. 8.15 – Exploded view of a rear brake assembly.
1 Brake back plate
2 Closing plug
3 Brake shoe hold-down pin
4 Brake shoe
5 Wheel brake cylinder
6 Brake shoe
7 Lower shoe return spring
8 Upper shoe return spring
9 Adjusting lever return spring
10 Adjusting lever
11 Adjuster strut with wheel
12 Brake drum
13 Adjuster pawl
14 Spring and spring seat

- Remove the brake shoe hold-down springs. Normally, a special tool is available, but a pair of pliers is suitable. Push with a finger from the rear of the brake back plate against the hold-down pin (10) and rotate the spring seat on the outside with a pair of pliers (1/4 of a turn) until the end of the pin can be guided through the slot in the spring seat. Remove the spring seat and the spring and withdraw the pin from the rear. Carry out the same operations on the other side of the vehicle.

- Lift the bottom end of the two brake shoes (7) and (9) of the brake back plate. Remove the return spring if still engaged in one of the shoes.
- Pull the two brake shoes at the upper ends apart and carefully disengage them from the wheel cylinder pistons. Place an elastic band around the wheel cylinder to keep the pistons in place. Otherwise, the wheel cylinder can be removed.
- Lift off the brake shoes and disengage the handbrake cable. To do this, compress the spring and disconnect the cable end as soon as possible.

Thoroughly clean all parts, including the back plate. If petrol is used, make sure that it cannot come near the wheel brake cylinder (if still fitted). The brake shoe linings can have a minimum thickness of 1.0 mm above the rivet heads, but if the thickness has worn down to 2.0 mm, we recommend to have the brake shoes replaced, to avoid the same work within a short time. Leaking wheel brake cylinders should be replaced.

8.3.2. INSTALLATION OF BRAKE SHOES

Place the brake shoes and other parts onto a bench as shown in Fig. 8.16 before commencing the installation.

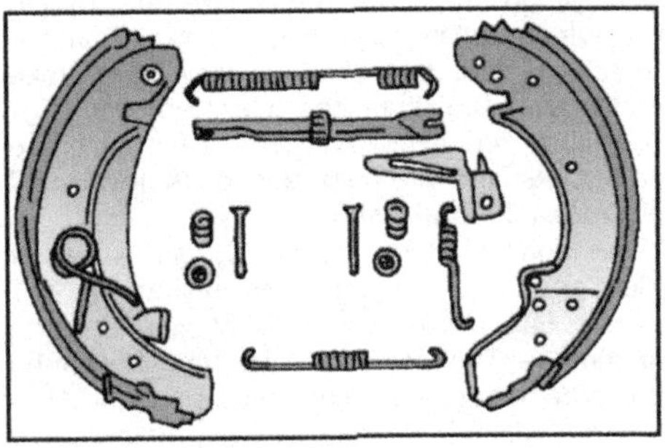

Fig. 8.16 – Arrange the rear brake parts on a bench as shown before installation.

- Refit the wheel brake cylinder to the brake back plate and connect the brake pipe at the rear (if removed).
- Fit the brake shoes to the brake back plate, engaging the handbrake cable with the rear show.
- Remove the elastic band from the wheel cylinder (if still fitted) and insert the upper shoe ends into the wheel cylinder pistons, without damaging the rubber caps. Place the lower ends against the abutment at the bottom of the brake back plate.
- Insert the brake shoe hold-down springs from the rear and place the spring and the spring seat over the pin from the front. Apply pressure to the pin from the rear, grip the spring seat with a pair of pliers and push it over the pin. As soon as the pinhead appears above the spring seat, turn the spring seat to lock the pin in position. Attach both brake shoes in the same manner.
- Insert the strut between the two brake shoes, but note that the strut with the R.H. thread is fitted to the L.H. brake assembly.
- Fit the upper and lower brake shoe springs between the brake shoes. First, engage the spring hook on one side and then stretch the spring with a pair of pliers, until the other end of the spring can be engaged. Make sure that the spring hooks are engaged as before dismantling.
- The next operation is carried out by referring to Fig. 8.17. Engage the spring (3) with the brake shoe and the other end with the adjuster lever (1).

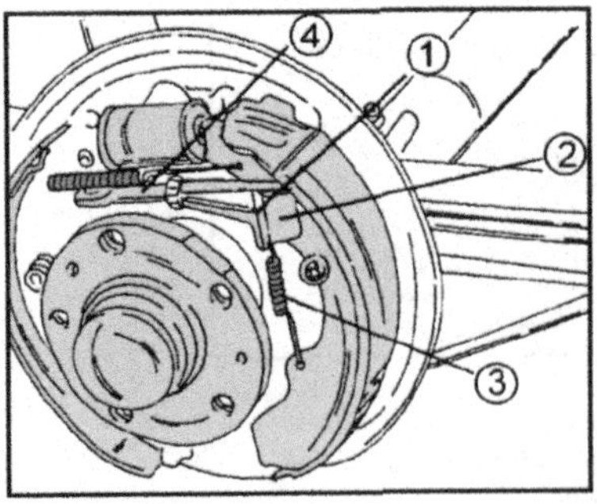

Fig. 8.17 – Details for the installation of the brake shoes.
1 Adjuster lever anchorage
2 Adjuster lever
3 Return spring
4 Adjuster strut with wheel

- Fit the adjuster lever (1). Fig. 8.18 shows more details of the installation. The end "a" of the lever must be placed under the link (2) and above the pin (3).

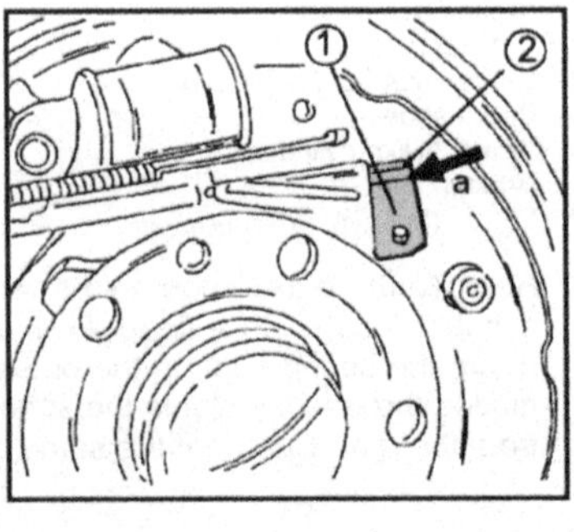

Fig. 8.18 – When fitting the adjuster lever (1), insert it at "a" underneath the strut (2).

- Centre the brake shoes on the brake back plate. Have a last look at the fitted brake shoes by comparing the assembly with Fig. 8.19, to make sure that all parts have been fitted correctly.
- Fit the brake drum and fit the two guide bolts (1.0 kgm/7.2 ft.lb.).
- Operate the hand brake and the foot brake a few times to set the self-adjusting mechanism into operation. This will also centre the brake shoes inside the drums.

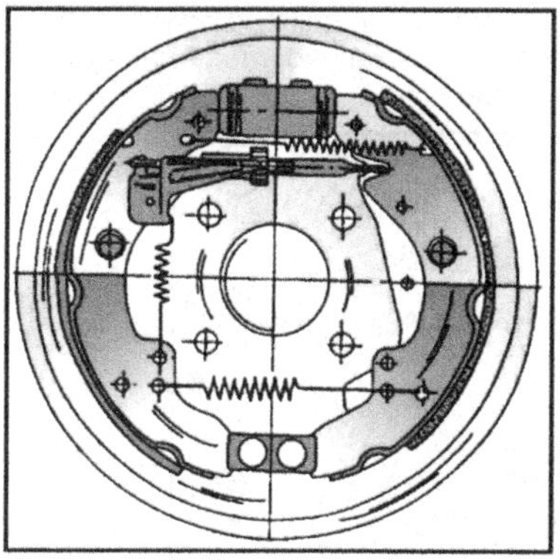

Fig. 8.19 – Before fitting the brake drums, compare the fitted brake shoes with the illustration.

- If the wheel cylinder has been removed, i.e., the brake pipe disconnected, bleed the brake system as described later on. Remember that each front caliper has two bleed screws (except models with ABS).
- Refit the wheels and lower the vehicle to the ground. Tighten the wheel bolts. Treat the brakes with the necessary care for the first few miles.

8.3.3. WHEEL BRAKE CYLINDERS

Wheel cylinders should not be overhauled. In the case of failure, fit a new cylinder (**Attention:** Different diameters). To remove a cylinder, remove the brake shoes as described above, disconnect the brake pipe at the rear of the back plate (union nut) and unscrew the cylinder securing bolts. A cylinder can also be removed with the brake shoes in position after unhooking the upper return spring(s) and pushing the brake shoes apart. In this case, you will have to be careful when the new cylinder is fitted, to avoid damage to the rubber caps.

Installation is a reversal of the removal procedure. Tighten the cylinder bolts to 2.0 kgm (14.5 ft.lb.) and the pipe union nut to 1.7 kgm (12 ft.lb.).

8.4. Brake Master Cylinder

Master cylinders of different diameters are used within the model range covered. Always make sure that the correct cylinder for the vehicle in question is fitted.

8.4.0. REMOVAL AND INSTALLATION

Note: The operation of the brake fluid level warning light can be checked by depressing the fluid reservoir cap (1) in Fig. 8.20. With the ignition switched on, the warning light in the dashboard must be illuminated.

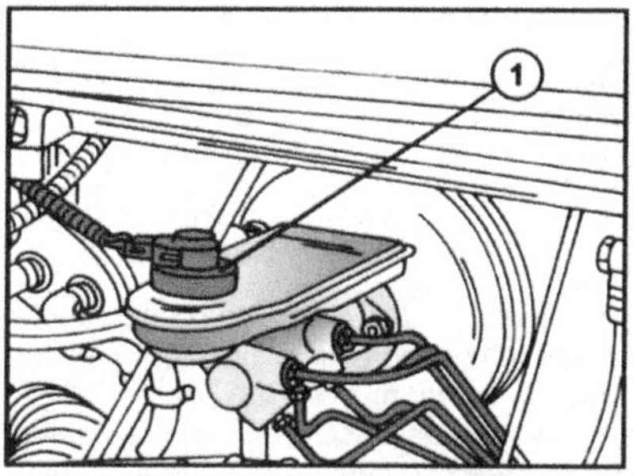

Fig. 8.20 – The operation of the warning light and the brake fluid level can be checked by pressing the reservoir cap (1).

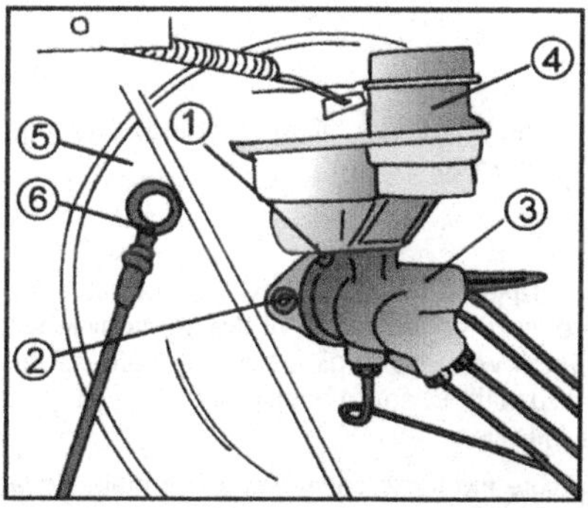

Fig. 8.21 – Master brake cylinder and brake servo unit.
1 Clip for reservoir
2 Securing nut for cylinder
3 Master brake cylinder
4 Reservoir
5 Brake servo
6 Vacuum hose

The removal and installation of the cylinder is a fairly easy operation. The cylinder is bolted to the front of the brake servo unit.

- Disconnect the battery earth cable.
- Remove the connector plug from the brake fluid level warning switch.
- Clean the area around the reservoir cap and the brake fluid pipes. Remove the reservoir cap and remove some of the brake fluids in the reservoir. A syringe can be used as shown in Fig. 8.22. Fig. 8.21 shows the fitted cylinder.

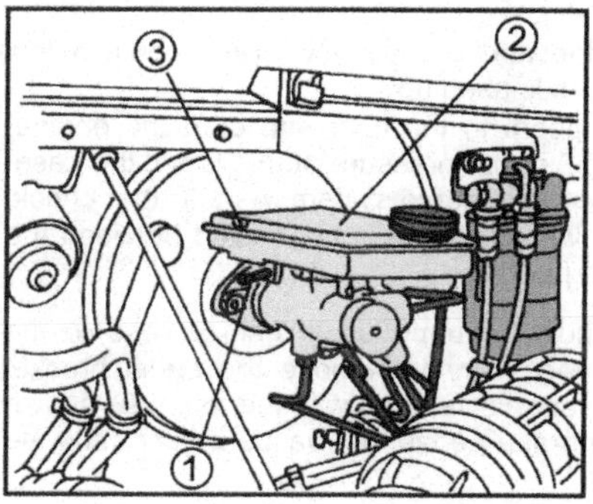

Fig. 8.22 – The securing clip (1) secures the reservoir (2) to the master brake cylinder (3).

- Refer to Fig. 8.22 and remove the securing clip (1) at the bottom of the fluid reservoir (2) and rock the reservoir to and fro until it is free of the rubber grommets in the cylinder (3). The reservoir is now taken out.
- Remove the brake pipe union nuts, withdraw the pipes and carefully bend the pipes to one side. Note where each pipe is connected and also take care not to spill brake fluid over the painted parts in the engine compartment.

- Remove the two nuts securing the cylinder to the front face of the brake servo unit and carefully lift out the cylinder. Take care not to drip brake fluid on painted surfaces. Hold a rag underneath the cylinder when lifting it out. Fig. 8.23 shows the details of the corresponding parts.

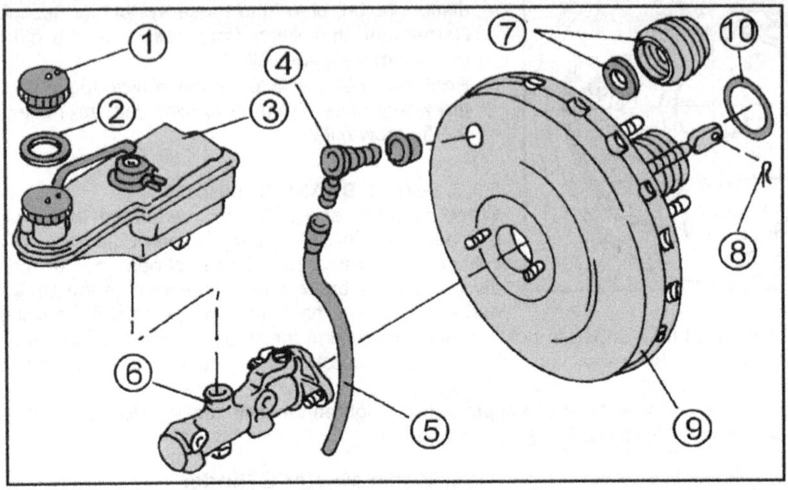

Fig. 8.23 – Master brake cylinder and brake servo unit.

1 Reservoir cap
2 Sealing ring
3 Fluid reservoir
4 Non-return valve
5 Vacuum hose (to pump)
6 Master brake cylinder
7 Rubber seal with filter
8 Clevis pin (clip)
9 Brake servo unit
10 Sealing ring

The installation of the cylinder is a reversal of the removal procedure.

The nuts are tightened to 2.0 kgm (14.5 ft.lb.). After installation of the cylinder, push the reservoir with the two connections into the rubber grommets of the cylinder body and secure it in position with the securing clip at the bottom. Make sure the clip is entered properly into its location.

Fill the fluid reservoir with clean brake fluid and bleed the brake system as described later on, noting the differences between vehicles with and without ABS. There are two bleed screws on the hydraulic ABS control unit.

8.4.1. CYLINDER REPAIRS

Master cylinders can be overhauled, but we recommend to fit a new cylinder, if the original one has developed a fault. Make sure a cylinder with the correct diameter is fitted.

8.5. Brake Servo Unit

Repair of the brake servo unit should not be attempted. Loss of the brake servo assistance will not influence the operation of the brake system, but you will find that the pedal pressure necessary to brake the vehicle will be substantially higher. Never allow the vehicle to free-wheel downhill with the engine switched off, as the vacuum for the brake servo unit will be used up very soon, leaving you without servo assistance.

Before the brake servo unit is removed, check its operation as follows.

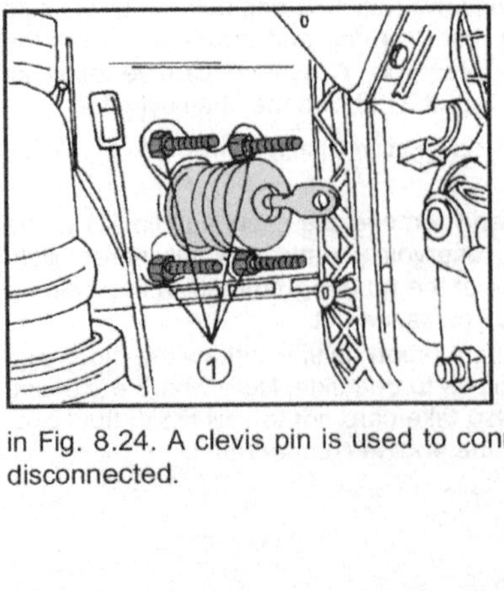

Fig. 8.24 – The nuts (1) secure the brake servo unit to the brake pedal bracket.

- Switch off the engine and operate the brake pedal a few times until the vacuum is used up.
- Depress the brake pedal to the floor and start the engine. The pedal must "sink" under pressure. If this is not the case, there is a fault in the brake servo system. As a further check, make sure that the vacuum hose is connected between the unit and the vacuum pump.

The master cylinder must be removed to gain access to the servo unit. It is also necessary to remove the pedal bracket inside the vehicle. The latter is secured with the four nuts shown in Fig. 8.24. A clevis pin is used to connect the push rod to the brake pedal. The vacuum hose must be disconnected.

Repairs are restricted to replacing the servo unit filter at the push rod end. To do this, push back the rubber boot, remove the retainer and remove the old filter with a pointed tool. The new filter must be cut with a pair of scissors from the outside towards the inside and then pushed over the push rod. Fig. 8.25 demonstrates the operation.

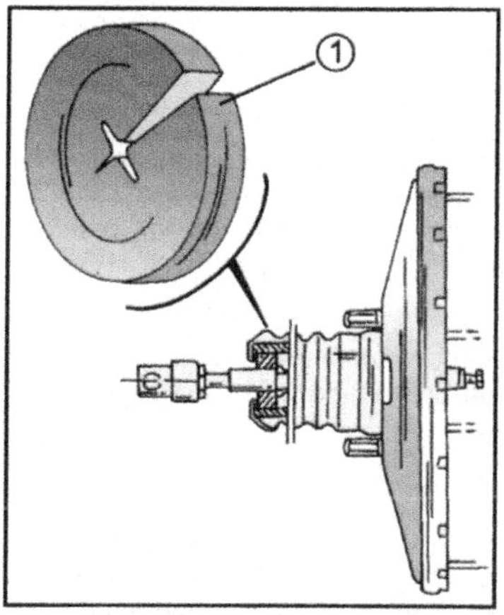

Fig. 8.25 – The filter (1) is located around the push rod in the end of the brake servo unit. Cut the filter as shown to fit a new one.

The check valve, i.e., non-return valve, can also be replaced. The location of the valve can be seen in Fig. 8.27. The valve is pushed into the servo unit and can be pulled out after disconnecting the vacuum hose. Check the condition of the rubber grommet, before the new valve is fitted.

If a new brake servo unit is fitted, adjust the length of the push rod. Details are shown in Fig. 8.27. The end of the push rod must be below the end face of the servo unit. The end of the push rod tip must be 0.1–0.3 mm below the end face, a very small clearance, but important for the correct operation of brake servo unit and master brake cylinder. Check by placing a steel ruler over the end face and measuring the clearance between end face and push rod tip. To adjust, slacken the locknut, rotate the push rod and re-tighten the locknut.

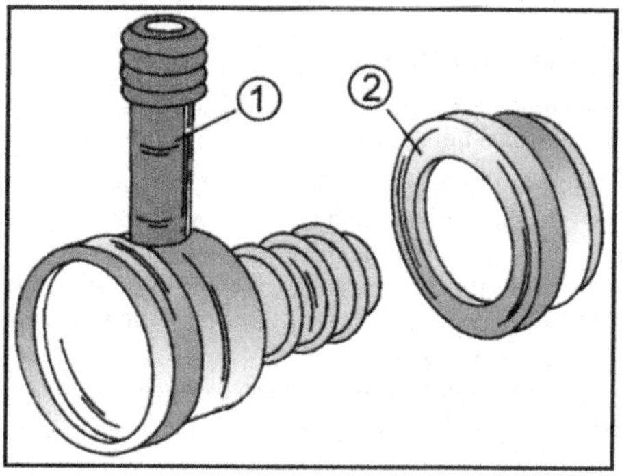

Fig. 8.26 – View of the non-return check valve in the brake servo unit. The vacuum hose is connected to the connector (1). The valve is pushed into the rubber grommet (2).

Fig. 8.27 – Adjusting the brake servo unit push rod. Measure from the end face to the tip of the push rod (1).

The securing nuts in Fig. 8.24 are tightened to 2.0 kgm (14.5 ft.lb.).

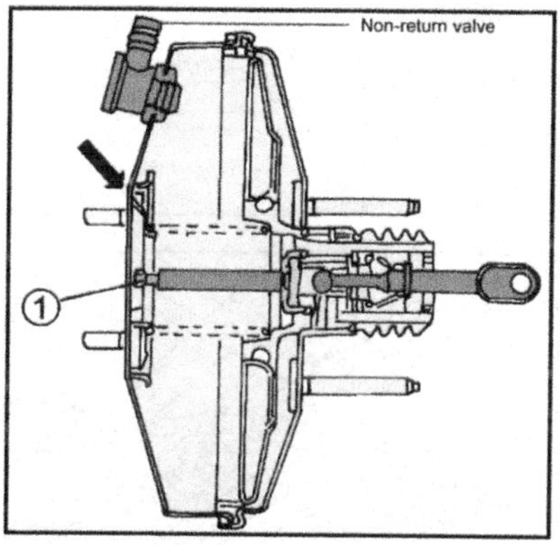

8.6. Bleeding the Brake System

Bleeding of the brake system is required, when the circuit has been opened or air has entered into the system, for example, if the fluid level in the reservoir has dropped too far. Before bleeding the system, clean the area around the reservoir filler cap and around the bleeding screws.

If only one-wheel brake cylinder or one brake caliper has been disconnected, it may be sufficient to bleed just the circuit connected to this brake unit, i.e., rear right and front left or rear left and front right, or the front axle and the rear axle, depending on the design of the brake circuit. Start the bleeding operation at the rear brakes and then the front brakes, if no ABS is fitted. The upper bleed screw of the front brakes is bled first. If ABS is fitted, proceed in the order rear right, front left, rear left and front right. You will need a helper, as the fluid level in the reservoir must be continuously observed, as it must not drop below the minimum level and, of course, to operate the brake pedal.

Important Note: Vehicles with ABS have two bleeding screws in the hydraulic control unit. These must be bled first. We must stress that you will need some experience with the bleeding of brake system to undertake the operations on such a vehicle.

Two bleed screws are fitted to the front brake calipers on vehicles without ABS. Vehicles with ABS only have one bleed screw.

Proceed as follows.

- Remove the rubber dust cap from the bleed screw and push a transparent bleeder hose over the bleed screw. Bleeder hoses are available from accessory shops. Hold the other end of the hose into a glass jar, filled with some brake fluid. Make sure that the bleeder hose remains at all times below the level of the brake fluid.
- Ask the helper to operate the brake pedal with slow strokes and to signal you, when the pedal is contacting the floor. Open the bleeder screw by half a turn. Fluid will run through the hose and into the jar. Close the bleed screw.
- Repeat the "pumping" operation and opening and closing of the bleed screw until brake fluid, free of air bubbles, is emerging into the glass jar. Remember to replenish the fluid reservoir in between.
- Repeat the bleeding operation in the given order on the other brake units. Finally, top-up the fluid reservoir to the required level. *Brake fluid drained out of the system into the glass jar must, however, not be used to fill the system.*

Final Note: The brake fluid should be changed every two years. Experience has shown that this necessity is normally overlooked. For your own safety, we strongly recommend it. Brake fluid is subject to heat and will deteriorate in service.

8.7. Adjusting the Handbrake

The rear wheels must be locked when the handbrake lever has been pulled by 1–5 notches. Place the rear of the vehicle on chassis stands. First, rotate the rear wheels to check whether the wheels can rotate freely. The handbrake requires only adjustment when the brake shoes or the handbrake cables have been replaced. If an adjustment is necessary, proceed as follows.

- Pull the handbrake to the third notch and jack up the rear end of the vehicle.

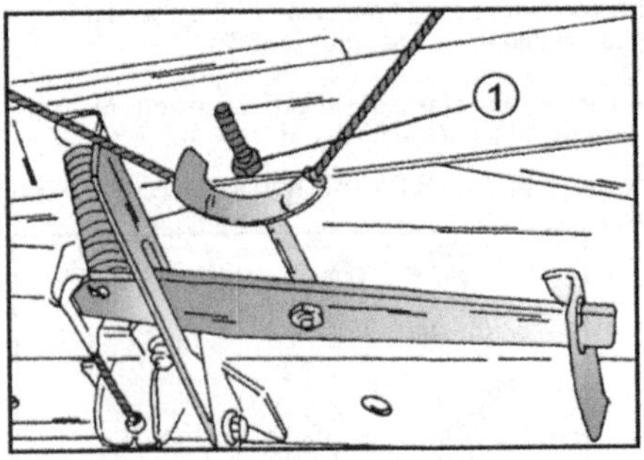

Fig. 8.28 – The nut (1) in the centre of the handbrake equaliser lever is used to adjust the handbrake cables.

- Tighten the nut (1) in Fig. 8.28 in the centre of the handbrake equaliser lever until the rear wheels can just be rotated under a fair effort. Re-tighten the locknut.
- Release the handbrake lever and then pull it once more to the third or fourth notch. Check that the rear wheels can no longer be rotated.
- Fully release the handbrake lever and check that both rear wheels can be rotated freely.
- Lower the vehicle to the ground.

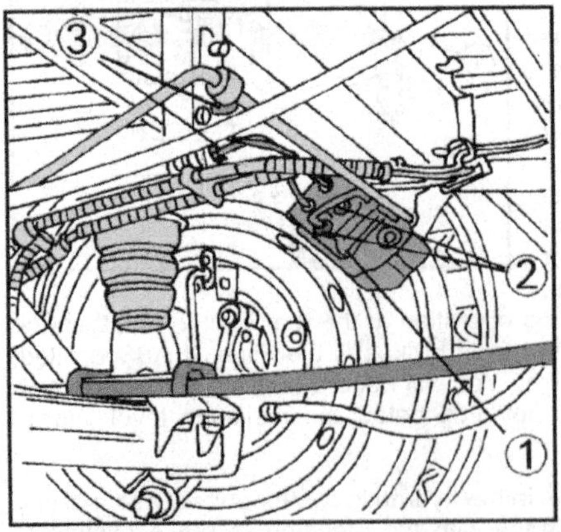

Fig. 8.29 – The location of the brake pressure regulator on the rear axle.

1 Brake pressure regulator
2 Securing screws
3 Pipe connections

8.8. Brake Pressure Regulator

The brake pressure regulator can only be adjusted with the help of two pressure gauges. Practically, every model in the covered range has a different pressure setting, depending on the fitted engine, the wheelbase, the carrying capacity and the type of body. Only your workshop will be able to check and/or adjust the regulator. The brake pressure regulator cannot be repaired and must be replaced if a fault has developed. The adjustment of the regulator takes place on the linkage, by shortening or lengthening it.

As the brake regulator must be adjusted, if it has been replaced, we recommend to entrust the operation to a workshop. Fig. 8.29 shows where the brake pressure regulator is fitted to the rear axle.

8.9. Vacuum Pump

Details of the vacuum pump have already been given in conjunction with the engine.

8.10. ABS

Models with the 2.0-litre engine have a Bosch ABS system. Additional parts in the system are the four-wheel speed sensors, one each in each of the wheels which are fitted to the steering knuckles at the front and the wheel hub carrier at the rear. The function of the sensors is to monitor the wheel speed by means of toothed rotors, which are fitted to the drive shafts (front) or axle stumps (rear). Normally, disc brakes are fitted to vehicles with ABS, but in this case, the drum brakes remain. Figs. 8.30 and 8.31 show where you will find these sensors on the vehicle.

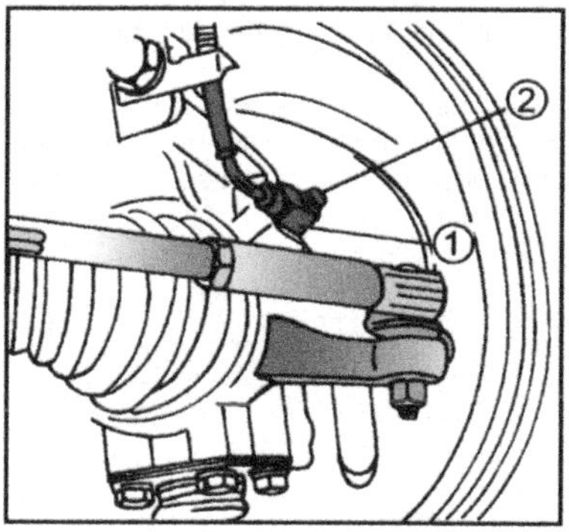

Fig. 8.30 – The location of the wheel speed sensor on the front suspension. The sensor (1) is secured by means of screw (2).

A computer receives the speed signals from the wheels and evaluates the values by comparing them with the pre-programmed values. By way of this process, the pressure to the wheel in question is reduced or increased by 4–10 times a second. Additionally, each wheel is controlled individually.

The Bosch system has 8 electrically operated valves and an electric pump. The pump is used to recover brake fluid, which has been re-directed during the pressure-reducing phase and builds up the pressure for the next operation of the ABS.

As the ABS provides you with additional safety, it is not our intention to describe any work on the system, with the exception of a few operations described below. A workshop is your best solution if you are not happy with the operation of the system. Also, note that certain precautions must be observed when bleeding the brake system.

8.10.1. Front Wheel Speed Sensors – Removal and Installation

- Jack up the front end of the vehicle and remove the wheel.

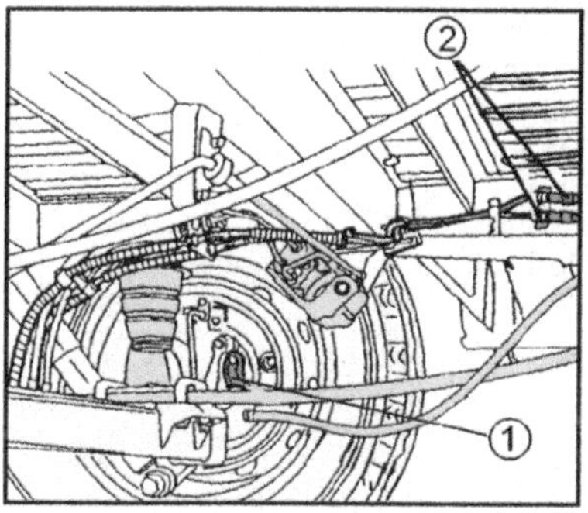

Fig. 8.31 – The location of the wheel speed sensor (1) on the rear axle. Make a note of the connection of cables (2).

- Remove the screw (2) in Fig. 8.30 and withdraw the wheel speed sensor out of the steering knuckle.
- Remove the cable from the attachment shown in the illustration and separate the cable connector. On the R.H. side, you will find the connector at the rear of the battery, and on the L.H. side below the fuel filler pipe.
- Clean the inside of the bore in the steering knuckle and insert the new speed sensor. The outside of the sensor should be greased with a little MP grease. Refit the screw and tighten it.
- Re-connect the plug and attach the cable.
- Fully move the wheels from lock to lock and check that the cable cannot touch other parts of the front suspension. Finally, refit the wheel and lower the vehicle to the ground.

8.10.2. Rear Wheel Speed Sensor – Removal and Installation

- Jack up the rear end of the vehicle and remove the rear wheel.
- Refer to Fig. 8.31 and remove the bolt out of the sensor (2) and the rear axle.
- Free the cables from the securing clip and separate the connector plugs. Mark the cables and plugs in accordance with the L.H. and R.H. sides to avoid re-connection the wrong way round.
- Withdraw the sensor out of the rear axle.
- Clean the inside of the bore in the axle and coat the new sensor with a little MP grease. Insert the sensor and fit and tighten the screw.
- Re-connect the connector plug, attach the cable and refit the wheel. Lower the vehicle to the ground.

8.10.3. Hydraulic Block and Electronic Control Unit

The bleeding of the system is not a straightforward operation and we, therefore, will not describe the removal and installation of the units in question. A new hydraulic block is pre-filled with brake fluid. If the electric pump can direct air into the unit, it will be very difficult or even impossible to bleed the hydraulic block, i.e., you may have to have it replaced – you have been warned.

Note: The workshop is using a bleeding apparatus to bleed the system. The primary circuit of the system is bled in the normal manner, in the order.

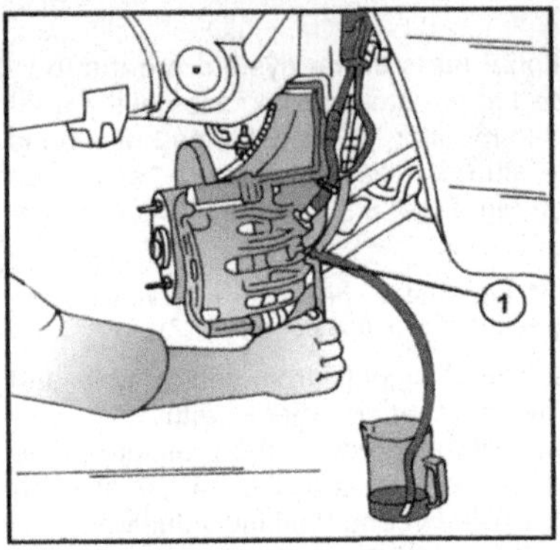

Fig. 8.32 – Bleeding a front caliper.

- The L.H. front wheel
- The R.H. front wheel
- The L.H. rear wheel
- The R.H. rear wheel

Bleeding can be carried out as described earlier on. Under no circumstances, switch on the ignition, or switch if off, if it is on, as this will lead to the operation of the electric pump and energise the electro-valves.

The secondary circuit must be bled if there is a leak in the secondary circuit, the ignition has been switched on (pump started) or the electro-valves have been supplied with brake fluid.

8.11. Tightening Torque Values

Bleeder screws	0.8 kgm (6 ft.lb.)
Brake pipe union nuts	1.4 kgm (10 ft.lb.)
Front brake disc securing screws	1.0 kgm (7.2 ft.lb.)
Caliper mounting bracket bolts	21 kgm (151 ft.lb.)
Caliper slide (guide) bolts	2.7 kgm (19 ft.lb.)
Rear wheel hub nuts	2.0 kgm (14.5 ft.lb.), adjust and then lock by peening
Front Drive Shaft Nuts:	
- Models 10Q and 14Q (10, 14)	45.0 kgm (324 ft.lb.)
- Model 18Q (Maxi)	50.0 kgm (360 ft.lb.)
Wheel Bolts:	
- Models 10Q and 14Q	16.0 kgm (115 ft.lb.)
- Model 18Q	18.0 kgm (130 ft.lb.)
Brake pipes, rear wheel cylinder	2.0 kgm (14.5 ft.lb.)
Brake pipes, master brake cylinder	1.5 kgm (10 ft.lb.)
Master cylinder securing nuts	2.0 kgm (14.5 ft.lb.)
Brake servo securing nuts	2.0 kgm (14.5 ft.lb.)

9. ELECTRICAL SYSTEM

9.0. Battery

Voltage .... 12 volts
Polarity .... Negative earth (ground)
Capacity .... 450 amps
Location .... Engine compartment

State of Charge:
Fully discharged: .... 1.120, specific gravity
Fully charged: .... 1.280, specific gravity

9.0.0. BATTERY CARE

The 12-volt battery consists of six cells, made up of positive and negative plates. Surrounded by a sulphuric acid solution. The battery provides the current to start the engine, for the glow plug system, the lighting of the vehicle and other current consumers.

The following maintenance operations should be carried out at regular intervals to extend the life of the battery and to always keep it at its peak performance.

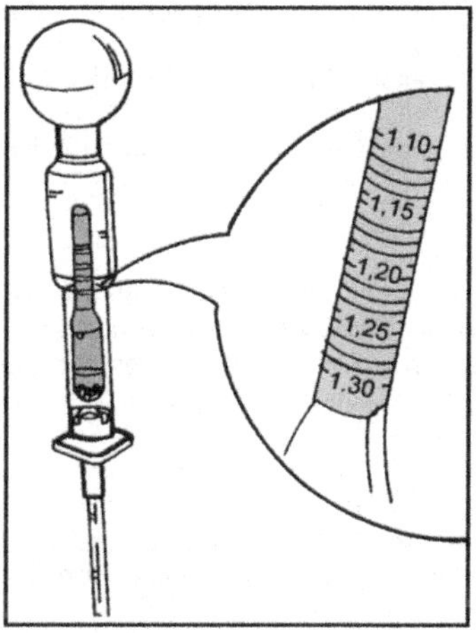

Fig. 9.1 – A hydrometer will show the specific gravity of the electrolyte by the position of the float inside the glass tube.

- Check the battery level once a week. If the battery is translucent, the level can be seen through the case. Otherwise, the filler plugs will have to be removed for inspection. If the electrolyte is below the separator plates, add distilled water (available from every petrol station). Do not over-fill the battery and wipe away any spilled water before replacing the filler plugs. Tap water must not be used to top-up the battery.
- Some batteries incorporate a battery checker, which enables an easy check of the battery condition. If a green spot can be seen, there is no need for further attention, i.e., the battery is sufficiently charged. If the battery checker turns dark, the battery must be re-charged. If the checker turns a light colour, electrolyte is missing in the battery.
- If frequent topping-up is necessary, it may be that the battery is over-charged by the alternator and the latter should be checked accordingly. A cracked battery case can also be the cause.
- The battery cables should always be firmly clamped, and the battery terminals must be free of corrosion to ensure good electrical conduct. Corroded areas can be cleaned with a soda solution and a wire brush. A thin coating of petroleum jelly should be smeared on battery posts before the cables are re-connected.
- Check the gravity of the electrolyte in each cell, using a hydrometer, as one is shown in Fig. 9.1. This is an indication of the charge condition of the battery. All cells should give the same reading. If there is a great variation in one cell, then either the electrolyte in the cell is weak due to being topped-up with distilled water or that cell is defective. In this case, a new battery must be fitted.
- The battery can be charged with a home charger, but follow the instructions of the manufacturer to avoid damage to the battery.
- Note that the radio security code must be known and **handy** after the battery is re-connected. All pre-programmed radio stations will also have been lost after the battery has been re-connected.

9.1. Alternator

9.1.0. ROUTINE PRECAUTIONS

The vehicles covered in this manual employ an alternator with an output of either 80 or 120 amps (also 95 amps), depending on the equipment. The alternator can be made by Bosch, Valeo or Mitsubishi. The alternator is driven by a belt from the crankshaft pulley, driving at the same time the compressor of an air conditioning system, if one is fitted. The alternator contains polarity-sensitive components, and the precautions below must be observed to avoid damage.

- Check the battery polarity before connecting the terminals. Immediate damage will result to the silicon diodes from a wrong connection – even if only momentarily.

- Never disconnect the battery or alternator terminals whilst the engine is running.
- Never allow the alternator to be rotated by the engine unless ALL connections are made.
- Disconnect the alternator multi-pin connector before using electric welding equipment anywhere on the vehicle.
- Disconnect the battery leads if a rapid battery charger is to be used.
- If an auxiliary battery is used to start the engine, take care that the polarity is correct. Do not disconnect the cables from the vehicle battery.
- When a new alternator is required, fit one as originally found in the vehicle.

9.1.1. SERVICING

Overhaul and checking of the alternator are best entrusted to a specialist or dealer who will have available the special equipment necessary. It is recommended that the owner should not use test instruments on the unit, unless he or she is familiar with the procedures and precautions necessary. Incorrect applications of meters and test equipment can quite easily cause damage to the diodes.

The alternator bearings are pre-lubricated and require no routine maintenance. The brushes can be inspected after a high mileage has been covered and can be replaced, if necessary. It should be noted, however, that the brush life can be expected to be quite long under normal conditions. Again, this is a job for a specialist shop.

9.1.2. ALTERNATOR MAINTENANCE

Periodically wipe away any dirt or grease that has accumulated on the outside of the alternator. If a water hose is used to clean the outside of the engine, cover the alternator to prevent water from entering. Check the security of the electrical cable connections on the alternator. The tension of the alternator drive belt should be checked at regulator intervals. Refer to Section 1.4.5.

Take extreme care when connecting the battery leads to the battery – positive to positive and negative to negative.

9.1.3. REMOVAL AND INSTALLATION

Before removal of the alternator, we must point out that the tension of the drive belt is adjusted by means of a checking instrument for the tension. We recommend that you read Section 1.4.5. before you attempt to remove the alternator. On models with air conditioning system, the alternator is driven together with the drive belt for the compressor. On models without air conditioning system, the alternator is driven together with the hydraulic steering pump.

- Disconnect the battery negative cable, followed by the positive cable.
- Note the location of the battery supply wires and disconnect them from the rear cover.
- Remove the drive belt, following the instructions in Section 1.4.5. and remove the alternator securing bolts. One bolt is at the upper end, and one bolt is at the lower end. Lift the alternator carefully from the vehicle.

The installation is a reversal of the removal procedure. Fit the drive belt as described in Section 1.4.5.

9.2. Starter Motor

9.2.0. TECHNICAL DATA

Make: .... Bosch (1.7 or 2.1 kW), Valeo (1.7 kW)
.... or Mitsubishi
Min. brush length: .... 13.0 mm (0.52 in.)

9.2.1. REMOVAL AND INSTALLATION

The starter motor can be removed from the clutch housing after the battery and the cables at the solenoid switch have been disconnected. The front of the vehicle should be placed on chassis stands to facilitate the access to the starter motor. The installation is a reversal of the removal procedure.

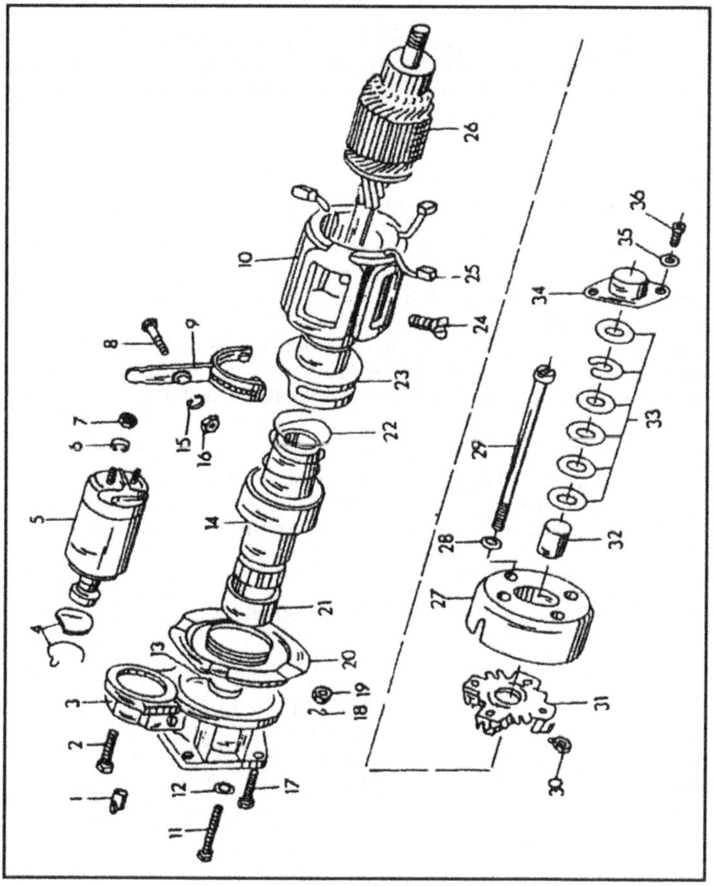

Fig. 9.2. Exploded view of a typical Bosch starter motor.

1 Starter motor bush
2 Starter motor bolt
3 Drive end bracket
4 Sundry parts
5 Solenoid switch
6 Spring washer
7 Nut, M8
8 Bolt, engagement lever
9 Engagement lever
10 Field coils
11 Bolt, M10
12 Plain washer
13 Stop ring
14 Starter motor drive
15 Spring washer
16 Spring washer
17 Bolt, M8
18 Spring washer
19 Nut, M8
20 Intermediate bracket
21 Engagement bush
22 Engagement spring
23 Engagement bush
24 Pole piece screw
25 Carbon brush
26 Armature
27 Commutator end
28 Plain washer
29 Through bolt
30 Brush spring
31 Brush holder
32 Commutator end bush
33 Sundry parts, com. end
34 Cover
35 Spring washer
36 Screw, M4

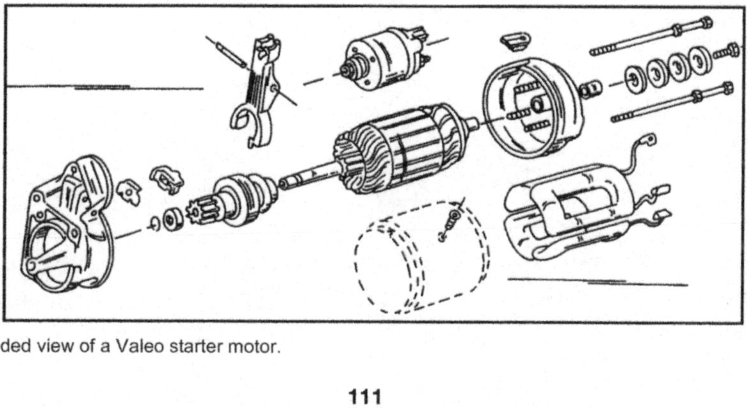

Fig. 9.3 – Exploded view of a Valeo starter motor.

9.3. Servicing

Starter motors are manufactured by Valeo, Bosch and also by Mitsubishi. We recommend to fit an exchange starter motor, if the original unit is no longer functioning. Various specialised companies are around and always have a starter motor for your specific model. There are, however, two exploded views of the starter motors used (Figs. 9.2 and 9.3).

9.4. Fuses

Different models use different fuses. Your Owner's Manual will give you the protected circuits for a particular model.

Before replacing a fuse, it is necessary to trace the cause of the fuse failure and to rectify it. To check whether a fuse is burnt out, inspect the metal strip across the top of the fuse. There should be no breaks in the centre. If the metal strip is interrupted, fit a fuse of the same value. Never fit a fuse with a smaller value to a position where a fuse with a higher value should be situated, or visa versa. Fuses are colour-coded, but again there are no particular guide lines. If a fuse is required, take the old fuse to the parts department of your dealer. This way you will be assured to obtain the correct fuse for the circuit in question. Never repair a fuse with a make-shift piece of metal across the fuse terminals.

9.5. Headlamps

As headlamps must be adjusted after one of them has been removed and the alignment must be checked optically, we refrain from giving any information on the removal and installation if the headlamps. We like, however, to point out that driving with misaligned headlamp beams is against the law.

9.6. Bulb Table

Headlamp main beams	H4, 55/60 watts
Fog lamps	H1
Brake/tail lamps	21/5 watts
Rear fog lamp	21 watts
Reversing light	21 watts
Indicator lights	5 watts
Number plate lights	5 watts
Front parking lights	5 watts

FAULT FINDING SECTION

The following section lists some of the more common faults that can develop in a motor car: both for petrol and diesel engines. For the purpose of this manual, references to diesel engines are, of course, first and foremost. The section is divided into various categories and it should be possible to locale faults or damage by referring to the assembly group of the vehicle in question. TDI engines require diagnostic instruments to locate faults.

The faults are listed in no particular order and their causes are given a number. By referring to this number, it is possible to read off the possible cause and to carry out the necessary remedies, if this is within the scope of your facilities.

ENGINE FAULTS

Engine will not crank:	1, 2, 3, 4
Engine cranks, but will not start:	5, 6, 7, 8
Engine cranks very slowly:	1, 2, 3
Engine starts, but cuts out:	5, 6, 9, 10
Engine misfires in the lower speed ranges:	5, 6, 9, 11
Engine misfires in the higher speed ranges:	5, 6, 11, 12
Continuous misfiring:	5, 6, 7, 10–15, 21, 22
Max. revs not obtained:	5, 6, 12, 22
Faulty idling:	5, 6, 8–11, 13, 15, 16, 21 and 22
Lack of power:	3, 5–11, 13–15, 22
Lack of acceleration:	5–8, 12, 14–16
Lack of max. speed:	5–8, 10, 12, 13–15, 22
Excessive fuel consumption:	3, 5, 6, 15, 16
Excessive oil consumption:	16–19
Low compression:	7, 11–13, 16, 20–22

CAUSES AND REMEDIES

1 Fault in the starter motor or its connection. Refer to "Electrical Faults".

2 Engine oil too thick. This can be caused by using the wrong oil, low temperatures or using oil not suitable for the prevailing climates. Depress the clutch whilst starting (models with manual transmission). Otherwise, refill the engine with the correct oil grade, suitable for diesel engines (for example, TOTAL DIESEL).

3 Moveable parts of the engine not run-in. This fault may be noticed when the engine has been overhauled. It may be possible to free the engine by adding oil to the fuel for a while.

4 Mechanical fault. This may be due to seizure of the piston(s), broken crankshaft, connecting rods, clutch or other moveable parts of the engine. The engine must be stripped for inspection.

5 Faults in the glow plug system. Refer to "Glow Plug Faults".

6 Faults in the fuel system. Refer to "Fuel Faults".

7 Incorrect valve timing. This will only be noticed after the engine has been reassembled after overhaul and the timing belt has been replaced incorrectly. Re-dismantle the engine and check the timing marks on the timing gear wheels.

8 Compression leak due to faulty closing of valves. See also under (7) or leakage past worn piston rings or pistons. Cylinder head gasket was blown.

9 Entry of air at inlet manifold, due to split manifold or damaged gasket.

10 Restriction in the exhaust system, due to damaged exhaust pipes, dirt in the end of exhaust pipe(s), kinked pipe(s), or collapsed silencer. Repair as necessary.

11 Worn valves or valve seats, no longer closing the valves properly. Top overhaul of engine is asked for.

12 Sticking valves due to excessive carbon deposits or weak valve springs. Top overhaul is asked for.

13 Cylinder head gasket was blown. Replace gasket and check block and head surfaces for distortion.

14 Camshaft worn, not opening or closing one of the valves properly, preventing proper combustion. Check and, if necessary, fit new camshaft.

15 Incorrect valve (tappet) clearance. There could be a fault in the hydraulic tappets.

16 Cylinder bores, pistons or piston rings worn. Overhaul is the only cure. Fault may be corrected for a while by adding "Piston Seal Liquid" into the cylinders, but will re-develop.

17 Worn valve guides and/or valve stems. Top overhaul is asked for.

18 Damaged valve stem seals. Top overhaul is asked for.

19 Leaking crankshaft oil seal, worn piston rings or pistons and worn cylinders. Correct as necessary.

20 Loose glow plugs, gas escaping past thread or plug sealing washer damaged. Correct.

21 Cracked cylinder or cylinder block. Dismantle, investigate and replace block, if necessary.

22 Broken, weak or collapsed valve spring(s). Top overhaul is asked for.

LUBRICATION SYSTEM FAULTS

The only problem the lubrication system should give is excessive oil consumption or low oil pressure, or the oil warning light not going off.

Excessive oil consumption can be caused by worn cylinder bores, pistons and/or piston rings, worn valve guides, worn valves stem seals or a damaged crankshaft oil seal or leaking gasket on any of the engine parts. In most cases, the engine must be dismantled to locate the fault.

Low oil pressure can be caused by a faulty oil pressure gauge, sender unit or wiring, a defective relief valve, low oil level, blocked oil pick-up pipe for the oil pump, worn oil pump or damaged main or big end bearings. In most cases, it is logical to check the oil level first. All other causes require the dismantling and repair of the engine. If the oil warning light stays on, switch off the engine IMMEDIATELY, as a delay could cause a complete seizure within minutes.

COOLING SYSTEM FAULTS

Common faults are: overheating, loss of coolant and slow warming-up of the engine.

Overheating:

- Lack of Coolant: Open the expansion tank cap with care to avoid injuries. Never pour cold water into an overheated engine. Wait until the engine cools down and pour in coolant whilst the engine is running.
- Radiator Core Obstructed by Leaves, Insects, etc: Blow with air line from the back of the radiator or with a water hose to clean.
- Cooling Fan Not Operating: Check fan for proper cut-in and cut-out temperature. If necessary, change the temperature switch or see your dealer.
- Thermostat Sticking: If sticking in the closed position, coolant can only circulate within the cylinder head or block. Remove thermostat and check as described in the “Cooling” section.
- Water Hose Split: Identified by rising steam from the engine compartment or the front of the vehicle. Slight splits can be repaired with insulation tape. Drive without expansion tank cap to keep the pressure in the system down, to the nearest service station.
- Water Pump Belt Torn: Replace and tension belt (if possible).
- Water Pump Inoperative: Replace the water pump.
- Cylinder Head Gasket Blown: Replace the cylinder head gasket.

Loss of Coolant:

1. Radiator Leaks: Using radiator sealing compound may stop slight leaks (follow the instructions of the manufacturer). In emergency, an egg can be cracked open and poured into the radiator filler neck.
2. Hose Leaks: See under 5, “Overheating”.
3. Water Pump Leaks: Check the gasket for proper sealing or replace the pump.

Long Warming-Up Periods:

1. Thermostat Sticking in the Open Position: Remove thermostat, check and, if necessary, replace.

DIESEL FUEL SYSTEM FAULTS

Engine is difficult to start or does not start	1–13
Engine starts, but stop soon afterwards:	14–20
Engine misfires continuously:	1–13
Bad idling:	14–20
Black, white or blue exhaust smoke:	21–29
Lack of power:	30–39
Excessive fuel consumption:	40–47

CAUSES AND REMEDIES

1. Fuel tank empty. Refuel.
2. Pre-glowing time too short. Operate until warning light goes “off”.
3. Cold starting device not operated. Pull cable and push in after approx. 1 mm.
4. Glow plug system inoperative (not applicable).
5. Electromagnetic cut-off device, loose or no current. Check cable to cut-off at top of injection pump. Ask a second person to operate ignition key and check if a “click” is heard. Either interrupted current supply or defective cut-off device.
6. Air in fuel system. Operate starter motor until fuel is delivered.
7. Fuel supply faulty. Slacken the injection pipes at injectors, and check if fuel is running out. Other faults: kinked, blocked or leaking injection pipes, blocked fuel filter and tank breathing system blocked. Wrong fuel for cold temperatures.

& Injection pipes refitted in wrong order alter repair.

9. Injection timing of pump out of phase: have the adjustment checked and corrected.
10. One or more injectors faulty, dirty or incorrect injection pressure. Have injectors repaired or replace them.
11. Injection pump not operating properly. Fit an exchange pump or have it repaired.
12. Valves not opening properly.
13. Compression pressures too low. See item "8" under "Engine Faults".
14. Idle speed not properly adjusted. Adjust.
15. Throttle cable not properly adjusted or sticking. Re-adjust or free-off.
16. Fuel hose between filter and pump not tightened properly. Tighten connections.
17. Rear mounting of injection pump loose or cracked. Tighten or replace.
18. See items 6, 7, 9, 11, 12 and 13.
19. Engine mounting not tightened properly or worn. Tighten or replace.
20. Sticking accelerator pedal. Free-off pedal.
21. Engine not at operating temperature. Check exhaust smoke colour again when engine is warm.
22. Too much acceleration at low revs. Use individual gears in accordance with acceleration.
23. Air cleaner contaminated. Clean or replace.
24. Fuel filter contaminated. Replace.
25. Max. speed adjustment incorrect. Re-adjust.
26. Injectors are dripping. Have them checked or replace faulty ones.
27. Injector nozzles sticking or broken. Replace injector.
28. Injection pressure too low. Have injectors checked and adjusted.
29. See items 9, 11, 12 and 13.
30. Throttle cable travel restricted. Re-adjust. Check that floor mats cannot obstruct pedal movement.
31. Throttle cable not correctly adjusted. Re-adjust.
32. Operating lever loose on pump. Re-tighten.
33. Max. speed not obtained. Re-adjust max. speed or have it adjusted.
34. Injector pipes restricted in diameter (near connections). Disconnect pipes and check that diameter is at least 2.0 mm (0.08 in.).
35. Heat protection sealing gaskets under injectors not sealing or damaged. Remove injectors and check. Replace if necessary. Fit the washers correctly.
36. Injection pressure of injectors wrong. Have them re-adjusted.
37. See items 6, 7, 9, 11 and 13.
38. See item 20.
39. See items 23, 24, 26 and 27.
40. Road wheels dragging. Brakes seized or wheel bearings not running freely.
41. Engine not running "free". Refers to new or overhauled engine.
42. Fuel system leaking. Check hoses, pipes, filter, injection pump, etc., for leaks.
43. Fuel return line blocked. Clean with compressed air if possible.
44. Idle speed too high. Re-adjust.
45. Max. speed too high. Re-adjust.
46. See items 10, 11, 12 and 13.
47. See items 24, 26, 27 and 28.

CLUTCH FAULTS

Clutch slipping:	1, 2, 3, 4, 5
Clutch will not disengage fully:	4, 6–12, 14
Whining from clutch when pedal is depressed:	13
Clutch judder:	1, 2, 7, 10–13
Clutch noise when idling:	2, 3
Clutch noise during engagement:	2

CAUSES AND REMEDIES

1. Insufficient clutch free play at pedal. Adjust in accordance with instructions in the "Clutch" section (if applicable).
2. Clutch disc linings worn, hardened, oiled-up, loose or broken. Disc distorted or hub loose. Clutch disc must be replaced.
3. Pressure plate faulty. Replace clutch.
4. Air in hydraulic system (only applicable to models with hydraulic clutch control). Low fluid level in clutch cylinder reservoir.
5. Insufficient play at clutch pedal and clutch release linkage (the latter in the case of mechanical operation). Adjust as described.
6. Excessive free play in release linkage (only for cable-operated clutch). Adjust or replace worn parts.
7. Misalignment of clutch housing. Very rare fault, but possible on transmissions with separate clutch housings. Re-align to correct.

8. Clutch disc hub binding on splines of main drive shaft (clutch shaft) due to dirt or burrs on splines. Remove clutch and clean and check splines.
9. Clutch disc linings loose or broken. Replace disc.
10. Pressure plate distorted. Replace clutch.
11. Clutch cover distorted. Replace clutch.
12. Fault in transmission or loose engine mountings.
13. Release bearing defective. Remove clutch and replace bearing.
14. A bent clutch release lever. Check lever and replace or straighten, if possible.

- The above faults and remedies are for hydraulic and mechanical clutch operation and should be read as applicable to the model in question, as the clutch fault finding section is written for all types of clutch operation.

STEERING FAULTS

Steering very heavy:	1–6
Steering very loose:	5, 7–9, 11–13
Steering wheel wobbles:	4, 5, 7–9, 11–16
Vehicle pulls to one side:	1, 4, 8, 10, 14–18
Steering wheel does not return to centre position:	1–6, 18
Abnormal tyre wear:	1, 4, 7–9, 14–19
Knocking noise in column:	6, 7, 11, 12

CAUSES AND REMEDIES

1. Tyre pressures not correct or uneven. Correct.
2. Lack of lubricant in rack and pinion steering.
3. Stiff steering linkage ball joints. Replace ball joints in question.
4. Incorrect steering wheel alignment. Correct as necessary.
5 Steering needs adjustment. See your dealer for advice.
6 Steering column bearings too tight or seized or steering column bent. Correct as necessary.
7. Steering linkage joints loose or worn. Check and replace joints as necessary.
8. Front wheel bearings worn, damaged or loose. Replace bearing.
9. Front suspension parts loose. Check and correct.
10. Wheel nuts loose. Re-tighten.
11. Steering wheel loose. Re-tighten nut.
12. Steering gear mounting loose. Check and tighten.
13. Steering gear worn. Replace the steering gear.
14. Steering track rods defective or loose.
15. Wheels not properly balanced or tyre pressures uneven. Correct pressures or balance wheels.
16. Suspension torsion bars (front) or springs (rear) weak or broken. Replace bar or spring in question or both.
17. Brakes are pulling to one side. See under "Brake Faults".
18. Suspension out of alignment. Have the complete suspension checked by a dealer.
19. Improper driving. We do not intend to tell you how to drive and are quite sure that this is not the cause of the fault.

BRAKE FAULTS

Brake Failure: Brake shoe linings or pads excessively worn, incorrect brake fluid (after overhaul), insufficient brake fluid, fluid leak, master cylinder defective, wheel cylinder or caliper failure. Remedies are obvious in each instance.

Brakes Ineffective: Shoe linings or pads worn, incorrect lining material or brake fluid, linings contaminated, fluid level low, air in brake system (bleed brakes), leak in pipes or cylinders and master cylinder defective. Remedies are obvious in each instance.

Brakes Pull to One Side: Shoes or linings worn, incorrect linings or pads, contaminated linings, drums or discs scored, fluid pipe blocked, unequal tyre pressures, brake back plate or caliper mounting loose, wheel bearings not properly adjusted and wheel cylinder seized. Rectify as necessary.

Brake Pedal Spongy: Air in hydraulic system. System must be bled of air.

Pedal Travel Too Far: Linings or pads worn, drums or discs scored, master cylinder or wheel cylinders' defective and system needs bleeding. Rectify as necessary.

Loss of Brake Pressure: Fluid leak, air in system, leak in master or wheel cylinders and brake servo not operating (vacuum hose disconnected or exhauster pump not operating). Place vehicle on dry ground and depress brake pedal. Check where fluid runs out and rectify as necessary.

Brakes Binding: Incorrect brake fluid (boiling), weak shoe return springs, basic brake adjustment incorrect (after fitting new rear shoes), piston in caliper of wheel cylinder seized, push rod play on master cylinder insufficient (compensation port obstructed) and handbrake adjusted too tightly. Rectify as necessary. Swelling of cylinder cups through the use of incorrect brake fluid could be another reason.

Handbrake Ineffective: Brake shoe linings worn, linings contaminated, operating lever on brake shoe seized, brake shoes or handbrake need adjustment. Rectify as necessary.

Excessive Pedal Pressure Required: Brake shoe linings or pads worn, linings or pads contaminated, brake servo vacuum hose (for brake servo) disconnected or wheel cylinders seized. Exhauster pump not operating (diesel engines). Rectify as necessary.

Brakes Squealing: Brake shoe linings or pads worn so far that metal is grinding against drum or disc. Inside of drum is full of lining dust. Remove and replace, or clean out the drum(s). Do not inhale brake dust.

Note: Any operation on the steering and brake systems must be carried out with the necessary care and attention. Always think of your safety and the safety of other road users. Make sure to use the correct fluid for the power-assisted steering and the correct brake fluid.

Faults in an ABS system should be investigated by a dealer.
Change the brake fluid in regular intervals (approx. every 2 years).

ELECTRICAL FAULTS

Starter motor failure:	2–5, 8, 9
No starter motor drive:	1–3, 5–7
Slow cranking speed:	1–3
Charge warning light remains on:	3, 10, 12
Charge warning light does not come on:	2, 3, 9, 11, 13
Headlamp failure:	2, 3, 11, 13, 14
Battery needs frequent topping-up:	11
Direction indicators not working properly:	2, 3, 9, 13, 14
Battery frequently discharged:	3, 10, 11, 12

CAUSES AND REMEDIES

1. Tight engine. Check and rectify.
2. Battery discharged or defective. Re-charge battery or replace if older than approx. 2 years.
3. Interrupted connection in circuit. Trace and rectify.
4. Starter motor pinion jammed in flywheel. Release.
5. Also 6, 7 and 8. Starter motor defective, no engagement in flywheel, pinion or flywheel worn or solenoid switch defective. Correct as necessary.
9. Ignition/starter switch inoperative. Replace.
10. Drive belt loose or broken. Adjust or replace.
11. Regulator defective. Adjust or replace.
12. Generator inoperative. Overhaul or replace.
13. Bulb burnt out. Replace bulb.
14. Flasher unit defective. Replace unit.

Wiring Diagrams

The following pages give the most popular wiring diagrams for the electrical components associated with the 2.0-litre HDi engine. There are some differences between model years 2001 and for models produced after 2002.

Wiring Diagram Index

Diagram 1	Starting and charging system (from 2002)
Diagram 2	Injection/Ignition system
Diagram 3	Engine cooling system (without A/C)
Diagram 4	Engine cooling system (with A/C)
Diagram 5	Front fog lamps
Diagram 6	Rear fog lamps
Diagram 7	Stop lamps (brake lights)
Diagram 8	Direction indicators, hazard warning lights
Diagram 9	Signal horn

Diagram 10	Side lamps
Diagram 11	Headlamps, dipped beams
Diagram 12	Headlamps, main beams
Diagram 13	Instrument panel lighting
Diagram 14	Interior lighting
Diagram 15	Interior lighting, 3 lamps
Diagram 16	Interior lighting, 4 lamps
Diagram 17	Windscreen washers
Diagram 18	Headlamp washer system
Diagram 19	Electrically operated front windows
Diagram 20	Electrically controlled mirrors
Diagram 21	Anti-lock brakes
Diagram 22	Heated rear screen and heated mirrors
Diagram 23	Starting and charging system – before 2002
Diagram 24	Pre-heating – before 2002
Diagram 25	Injection/Ignition – before 2002
Diagram 26	Engine cooling system – before 2002
Diagram 27	Coolant temperature warning
Diagram 28	Water in diesel fuel filter warning
Diagram 29	Engine oil pressure and oil level
Diagram 30	Handbrake warning lamp
Diagram 31	Headlamps – before 2002

Please Note: The identification numbers for the electrical components are given as available.

WIRING DIAGRAM LEGEND

Not all current consumers are fitted to all models.

BB00	Battery
BB10	Supply box
BF00	Fuse box
BMF1	Max. fuse box
BMF2	Max. fuse box
CA00	Ignition switch
C1300	Engine test connector
C4640	Tachograph fuse Connector
C7000	ABS diagnostic connector
MF175	Fuse, 175 A
PS00	Connector board, 4 × 4 system
V1000	Charge warning light
V1150	Pre-heating warning lamp
V1300	Engine diagnostic warning lamp
V2000	Rear fog lamp warning lamp
V2300	Warning lamp, hazard warning lights
V3310	Warning lights, direction indicators
V2600	Warning light, side lamps
V2620	Main beam warning lamp
V2660	Warning light, front fog lamps
V4010	Warning light, coolant level
V4020	Warning light, coolant temperature
V4050	Warning light, water in fuel
V4110	Oil pressure warning Lamp
V5300	Warning lamp, low fuel
V4420	Warning light, hand-brake and brake fluid, also brake pad wear
V6235	Warning light, anti-theft device
V7000	Warning light, ABS
V8110	Warning light, heated rear screen
0004	Instrument cluster
1005	Starter inhibitor switch
1010	Starter motor
1020	Alternator
1021	General support relay
1041	Emergency stop switch
1042	General relay
1043	Re-arming push button
1044	Diode unit
1104	Electro-valve, advance Corrector
1150	Pre-heating control unit
1156	Post-heating relay
1157	Post-heating thermal switch
1160	Glow plugs
1203	Inertia switch
1204	Impact safety relay
1208	Diesel injection pump
1211	Fuel gauge
1220	coolant temperature sensor
1252	Advance corrector Relay
1254	Load lever switch
1255	Fuel pump cut-out electro-valve
1316	Engine speed sensor
1331	Injector No. 1
1332	Injector No. 2
1333	Injector No. 3
1334	Injector No. 4
1400	TDC sensor
1500	Cooling fan relay
1505	Cooling fan thermal switch
1506	Cooling fan resistor
1510	Cooling fan
1526	Cooling fan timer
1620	Vehicle speed sensor
2000	Rear fog lamp switch
2001	Light/wiper switch
2100	Stop lamp switch
2200	Reversing lamp switch
2300	Hazard warning light switch
2305	Flasher unit
2320	Front L.H. flasher
2325	Front R.H. flasher
2340	L.H. side flasher
2345	R.H. side flasher
2520	Horn
2600	Lamp rotator
2605	Dipped beam relay
2606	Main beam relay
2610	L.H. headlamp
2615	R.H. headlamp
2630	Rear L.H. lamps on bodyshell
2633	R.H. number plate light
2634	Rear R.H. lamps on bodyshell
2635	L.H. number plate light
3000	Front L.H. door aperture switch
3001	Front R.H. door aperture switch
3010	Front Interior lamp
3019	Interior lamp switch
3020	Rear interior lamp
3029	Central interior lamp switch
3030	Central interior lamp
3086	Blue interior lamp switch
3087	Parking lamps switch
3088	Parking lamps relay
4025	Engine coolant thermal switch and temperature sensor
4026	Engine coolant temperature switch
4050	Water in fuel sensor
4110	Engine oil pressure switch
4310	Fuel gauge
4311	Fuel gauge stabiliser
4315	Fuel gauge rheostat
4400	Handbrake switch
4410	Brake fluid level switch
4430	L.H. brake pad wear indicator contact
4431	R.H. brake pad wear indicator contact
4640	Tachograph
5015	Wiper motor
5100	Pump, windscreen washer
5405	Pump, headlamp washer

5410	L.H. headlamp washer motor
5415	R.H. headlamp washer motor
6004	Window switch, driver side
6005	Window switch, passenger side
6021	Window relay
6040	Window motor, driver side
6235	Door locking control unit
6240	L.H. door locking motor
6245	R.H. door locking motor
6255	R.H. side door locking motor
6260	Rear door locking motor
6406	Rear view mirror switch
6410	Driver's door mirror
6415	Passenger door mirror
6540	Seat belt unit
6700	Control switch, differential lock
6701	Electro-valve, differential lock
6702	4 × 4 selector valve
6703	4 × 4 selector valve 4 × 4 selector relay
6706	4 × 4 selector switch
6712	Differential unlocking switch
6740	Differential locking relay
7000	Front L.H. ABS sensor
7005	Front R.H. ABS sensor
7010	Rear L.H. ABS sensor
7015	Rear R.H. ABS sensor
7020	ABS control unit
7025	Brake pressure unit
7029	ABS pump fuse
7030	Electric pump unit
7020	Clock
8000	A/C switch
8005	Slow-speed air blower relay
8006	Evaporator thermal sensor
8007	Pressostat
8020	A/C compressor
8040	Air blower speed control
8041	Additional heating switch
6046	Air blower speed resistor
8048	Additional heating relay
8049	Air blower motor
8060	Additional air blower motor
8061	A/C unit relay
8100	Front cigar lighter
8110	Heated rear window switch
8115	Heated rear window relay
8120	Heated rear window
8121	Control module transponder
8122	Digital electro-valve
8310	Heated rear seat
8410	Radio
8420	L.H. loudspeaker
8425	R.H. loudspeaker

Wiring Looms – Abbreviations

AVID	Wiring loom, front right
FIAF	Wiring loom, front panel
ES/P	Wiring loom, headlamp washer
ABR	Wiring loom, ABS
ABR/PB	Wiring loom, ABS, dashboard
CSL	Wiring loom, 4 × 4 operation
P/B/C	Wiring loom, add. dashboard
PLAF	Wiring loom, interior lights
CLM	Wiring loom, A/C system
PR AV/G	Wiring loom, L.H. front door
PRAR	Wiring loom, rear door
PR AV/D	Wiring loom, R.H. front door
LI PR	Wiring loom, door connection harness
AR/D	Wiring loom, R.H. rear
INT/F/AV G	Front L.H. door switch
CHR/MOT	Wiring loom, tachograph
AVIG	Wiring loom, front left
GMV	Wiring loom, cooling fan
MOT	Wiring loom, engine
ABR/AV	Wiring loom, ABS, front
S/CH	Wiring loom, seat heater
PLAF/C	Wiring loom, Interior lamp
P/B	Wiring loom, dashboard
PLAF/AR	Wiring loom, rear interior lamp
CL/AD	Wiring loom, add. Heater
PR AR/G	Wiring loom, L.H. rear door
PRAV/I/D	Wiring loom, 2nd front door
PR AR/D	Wiring loom, R.H. rear door
AR/G	Wiring loom, L.H. rear
AR/SP	Wiring loom, rear under floor
INT/F/AF D	Front R.H. door switch
CHR/PB	Wiring loom, dashboard, tachograph

Cable Colour Code

BA = white
BE = blue
BG = beige
GR = grey
JN = yellow
MR = brown
MV = Mauve
NR = black
OR = orange
RG = red
RS = pink
VE = green
VI = violet
AY = light blue

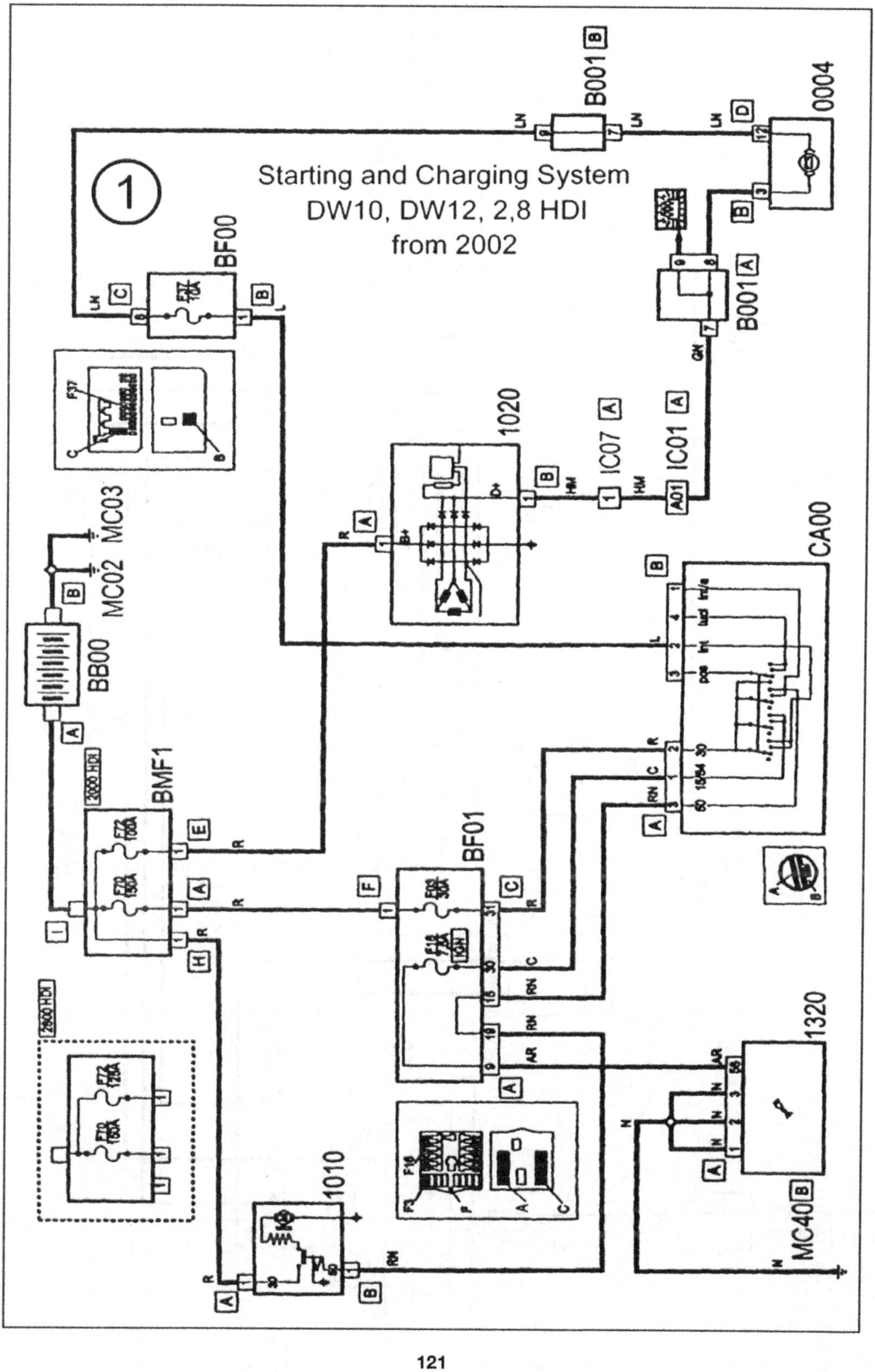
1
Starting and Charging System
DW10, DW12, 2,8 HDI
from 2002
B001 B
0004
BF00
B001 A
1020
IC07 A
IC01 A
MC02 MC03
BB00
CA00
BMF1
BF01
1320
1010
MC40 B
2800 HDI
3000 HDI

Injection-Ignition
DW10-DW12
from 2002

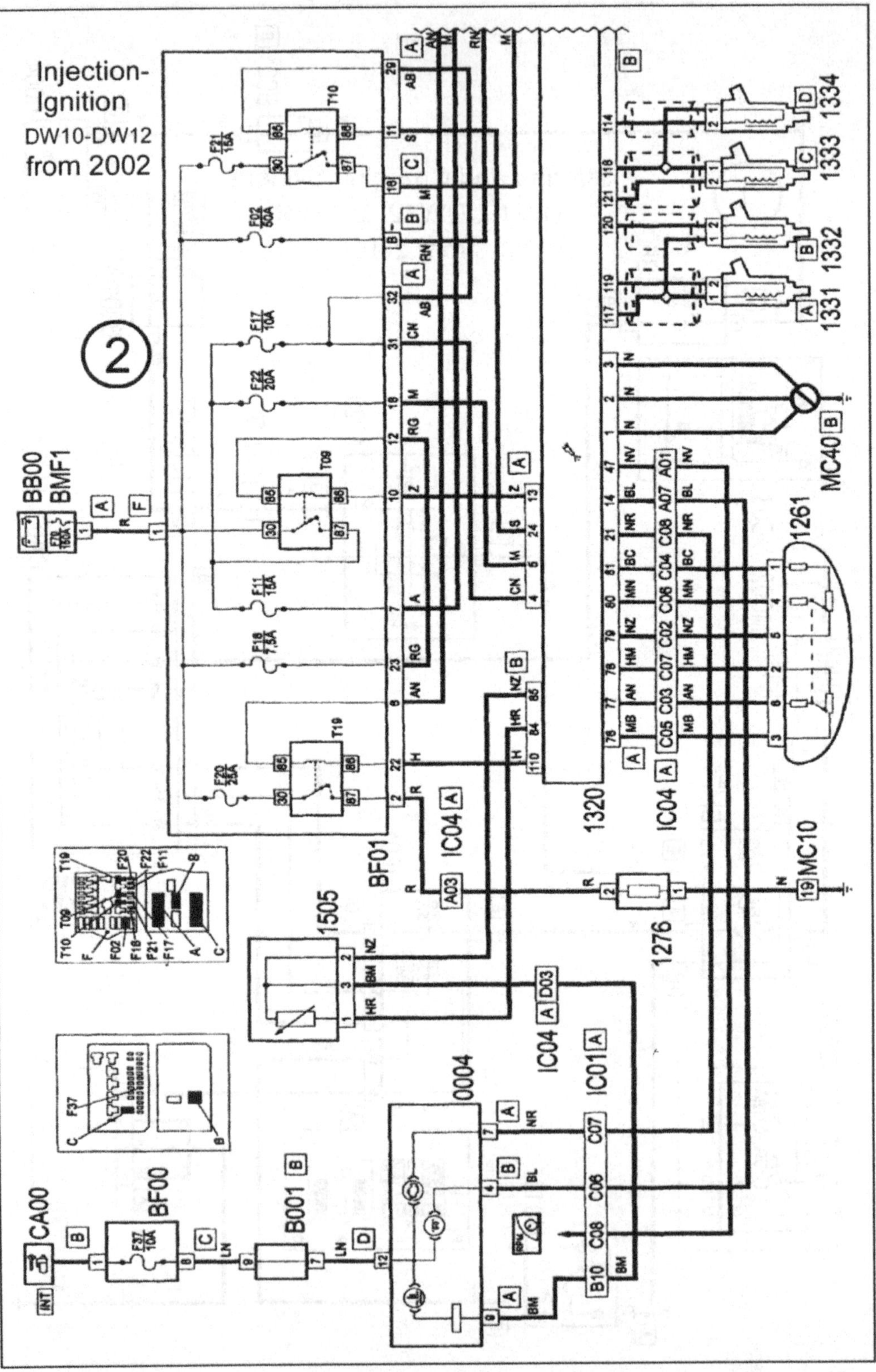

3 Cooling System without A/C System

DW10-DW12

from 2002

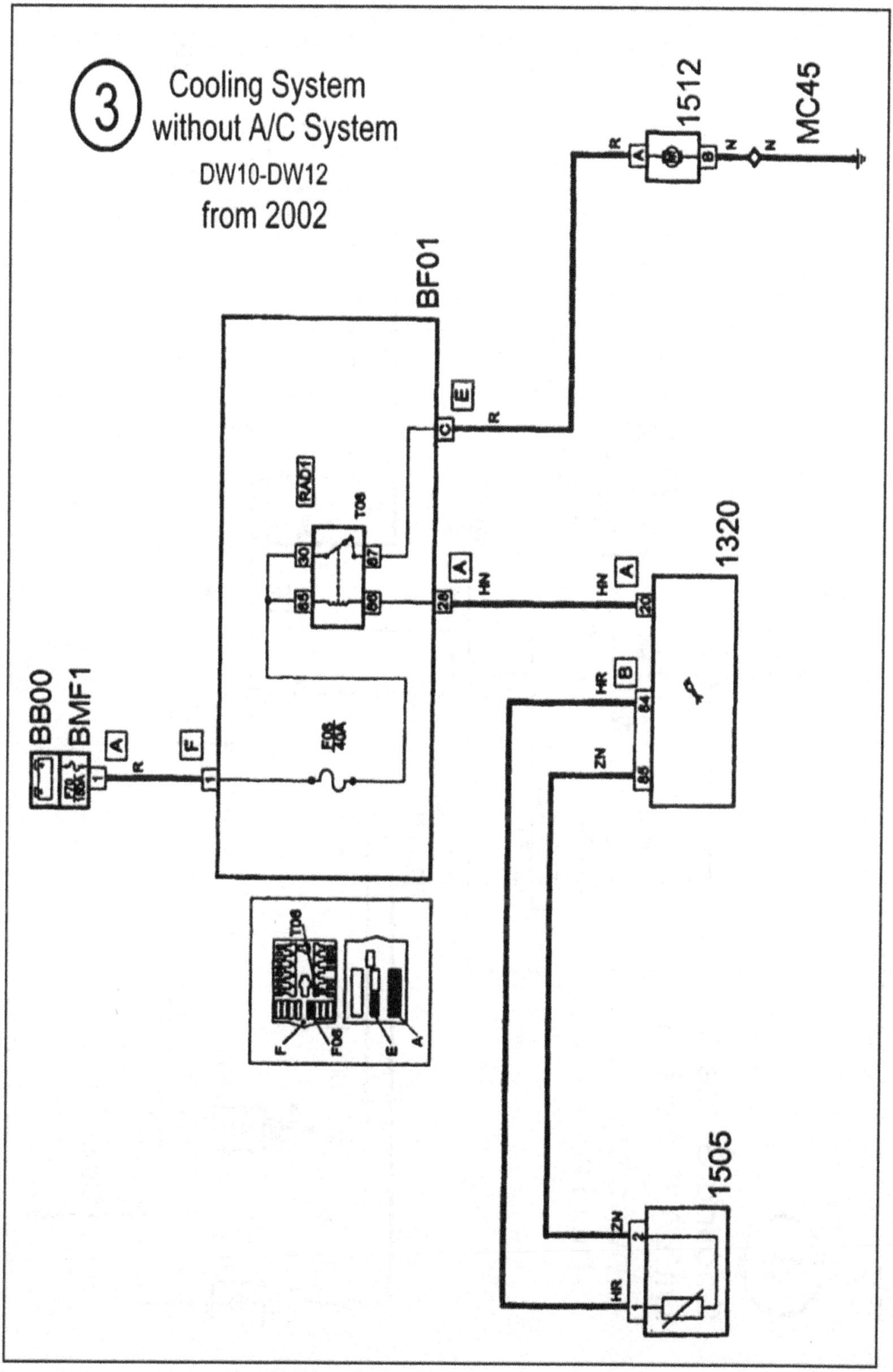

4

Cooling System - with A/C System

DW10-DW12
from 2002

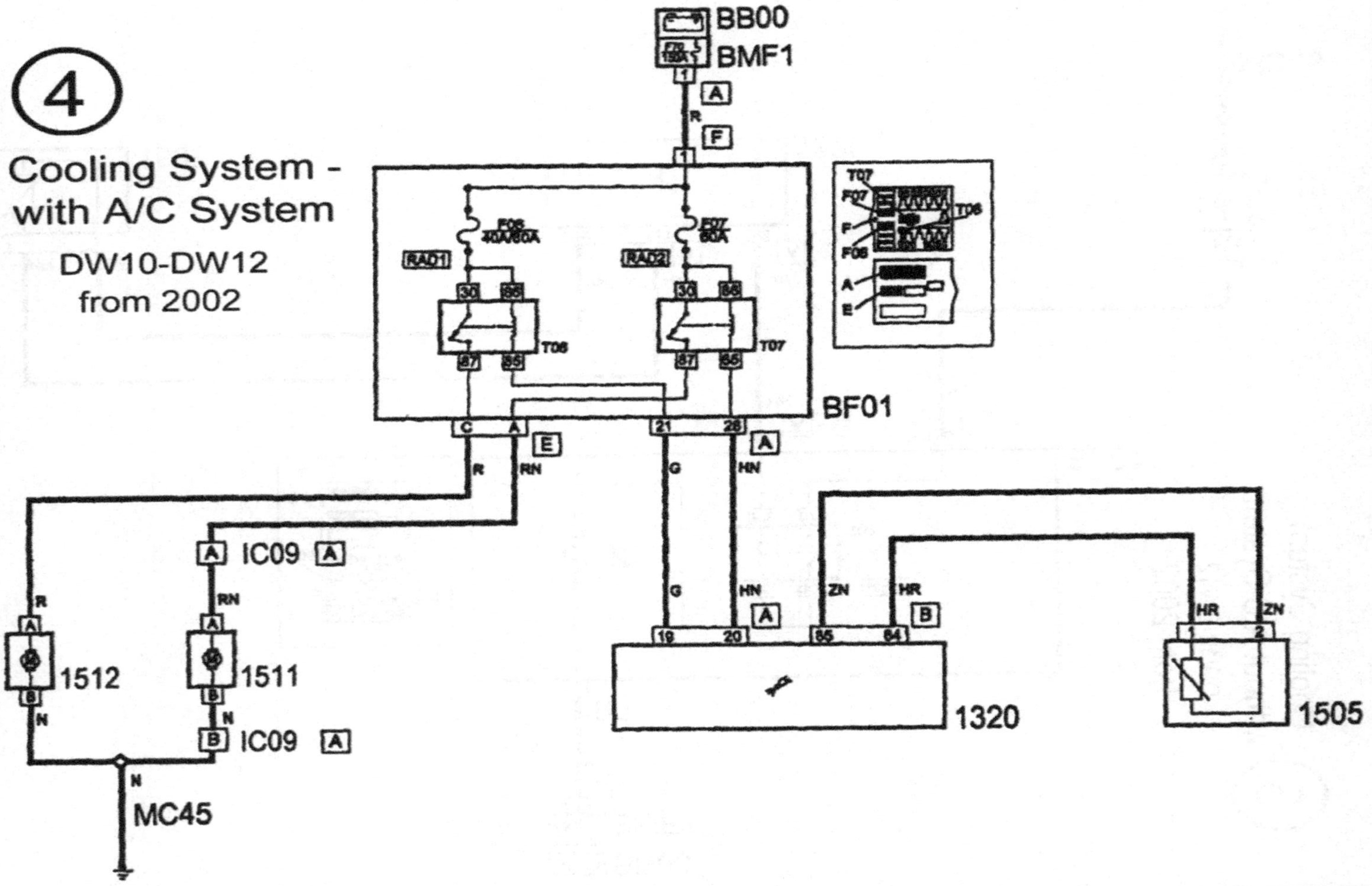

5 Front Fog Lamps

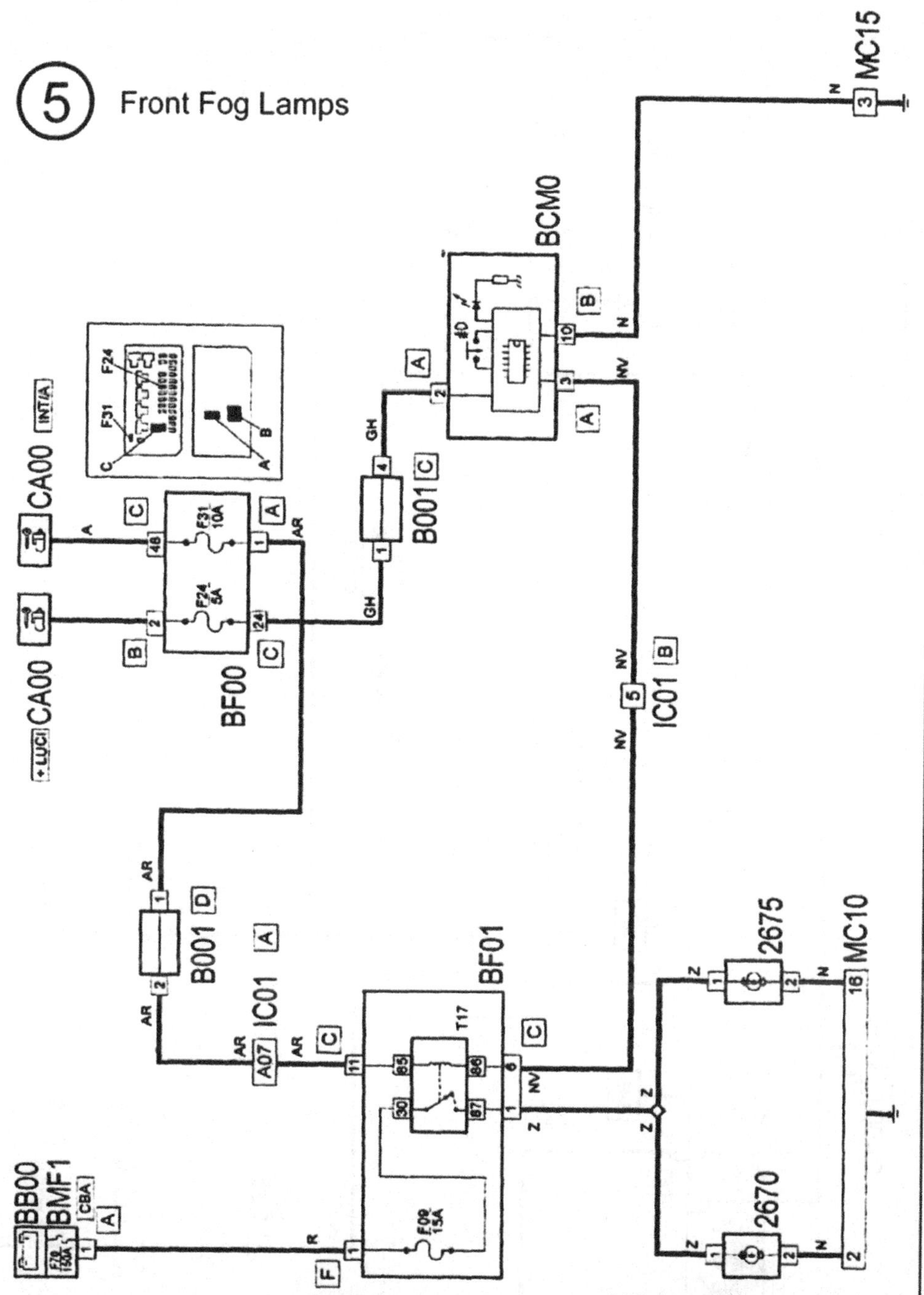

Wiring Diagrams
5 to 22 from
2002
(Relay 2)
from Chassis No. 7000001

Rear Fog Lamps

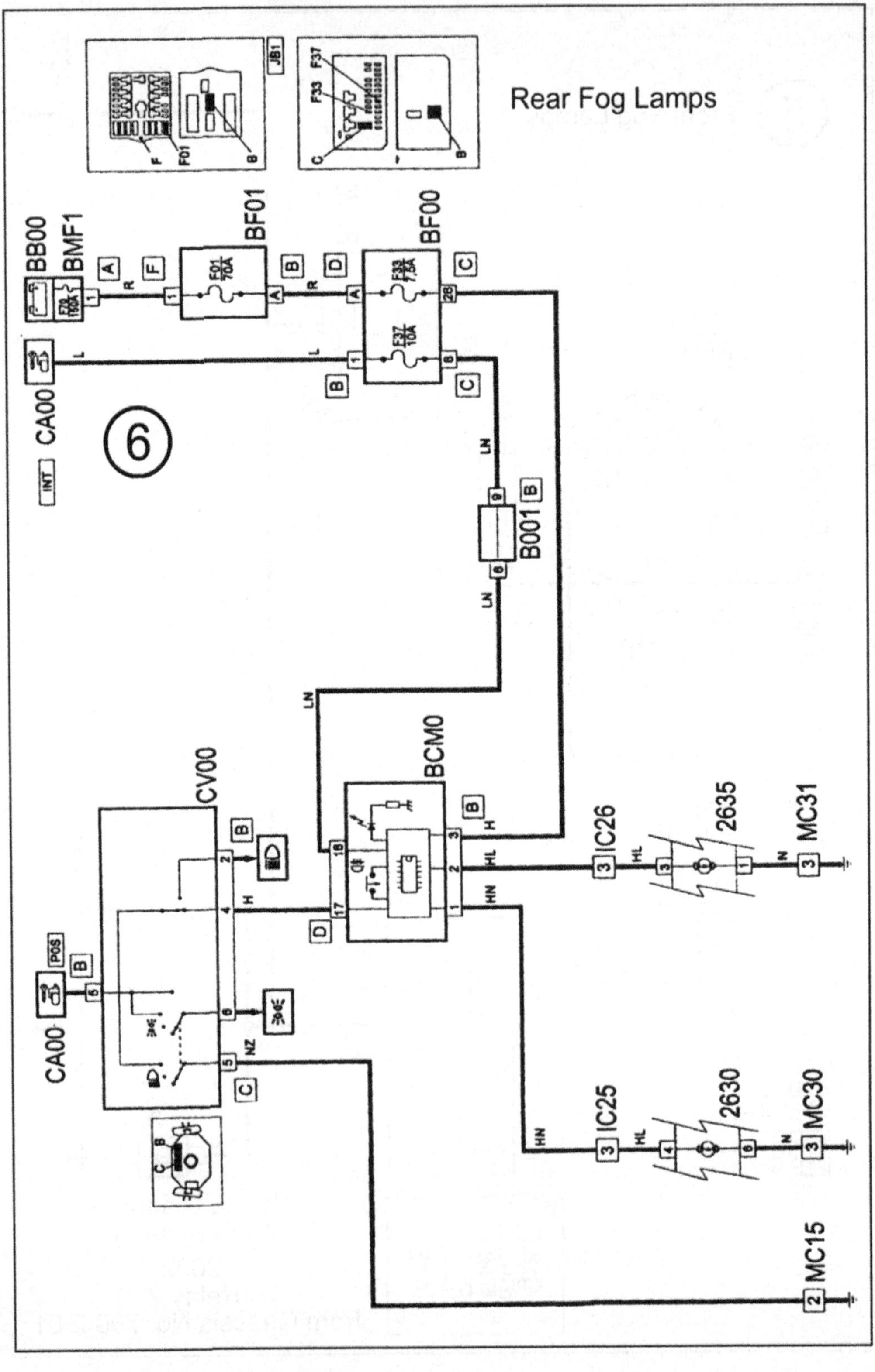

7 Stop Lamps

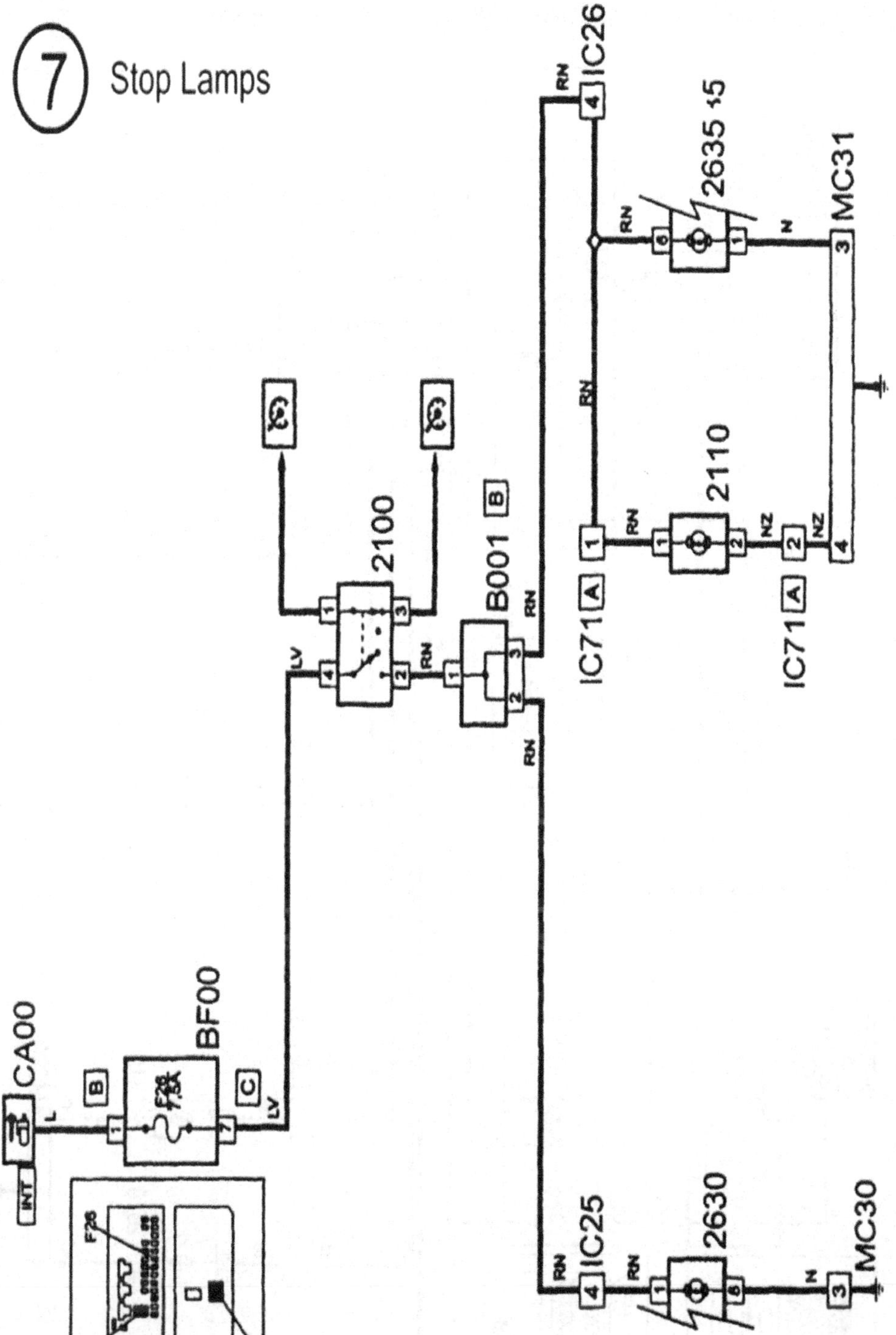

INT
CA00
L
B
1
F26
7,5A
BF00
7
C
LV
LV
4
1
2100
2
3
RN
1
B001
B
2
3
RN
RN
RN
4
IC25
RN
1
2630
8
N
3
MC30
IC71
A
1
RN
RN
4
IC26
RN
1
2110
2
NZ
IC71
A
2
NZ
4
RN
6
2635 45
1
N
3
MC31
C
F26
B

8 Direction Indicators - Hazard warning light

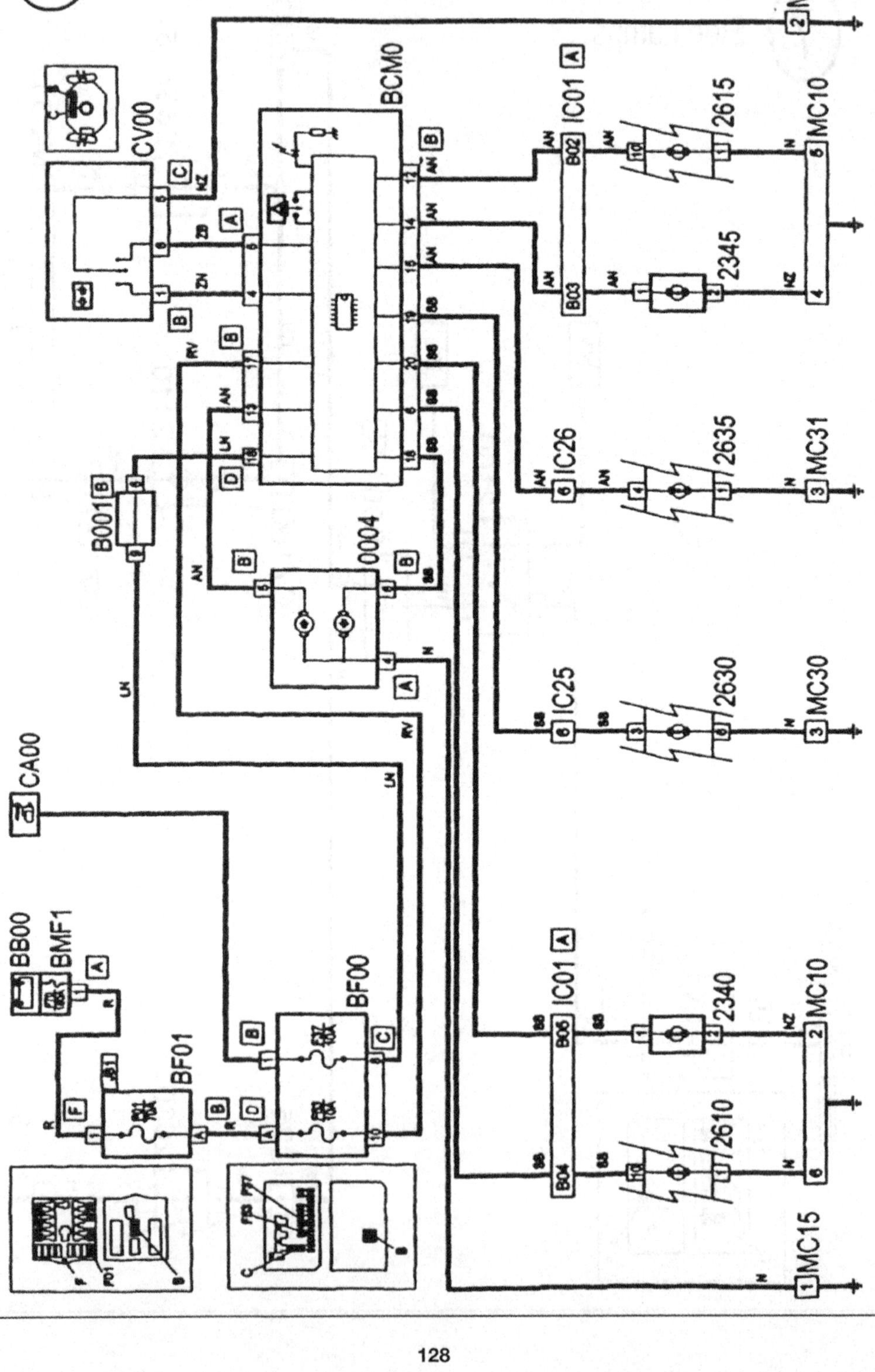

9 Signal Horn

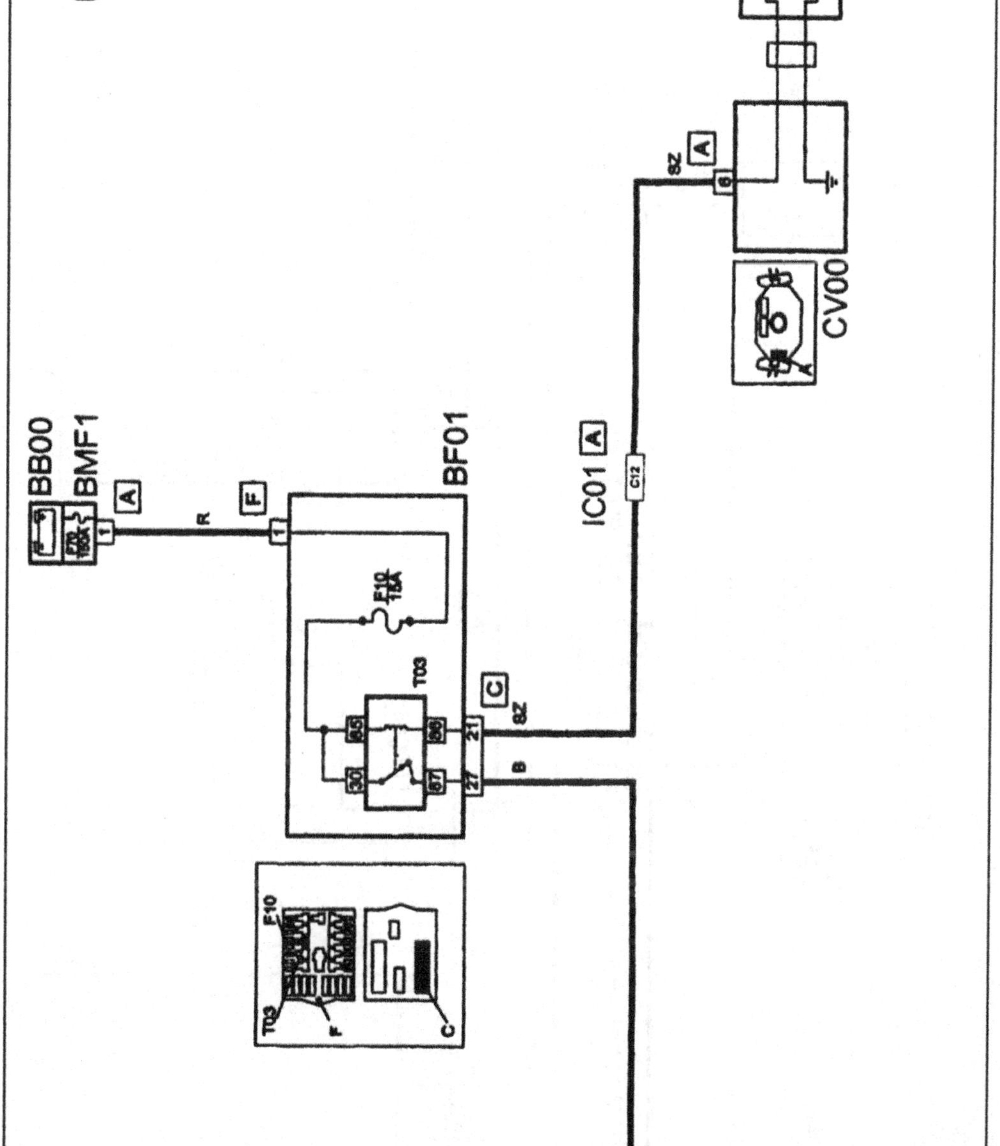

(10) Side Lamps

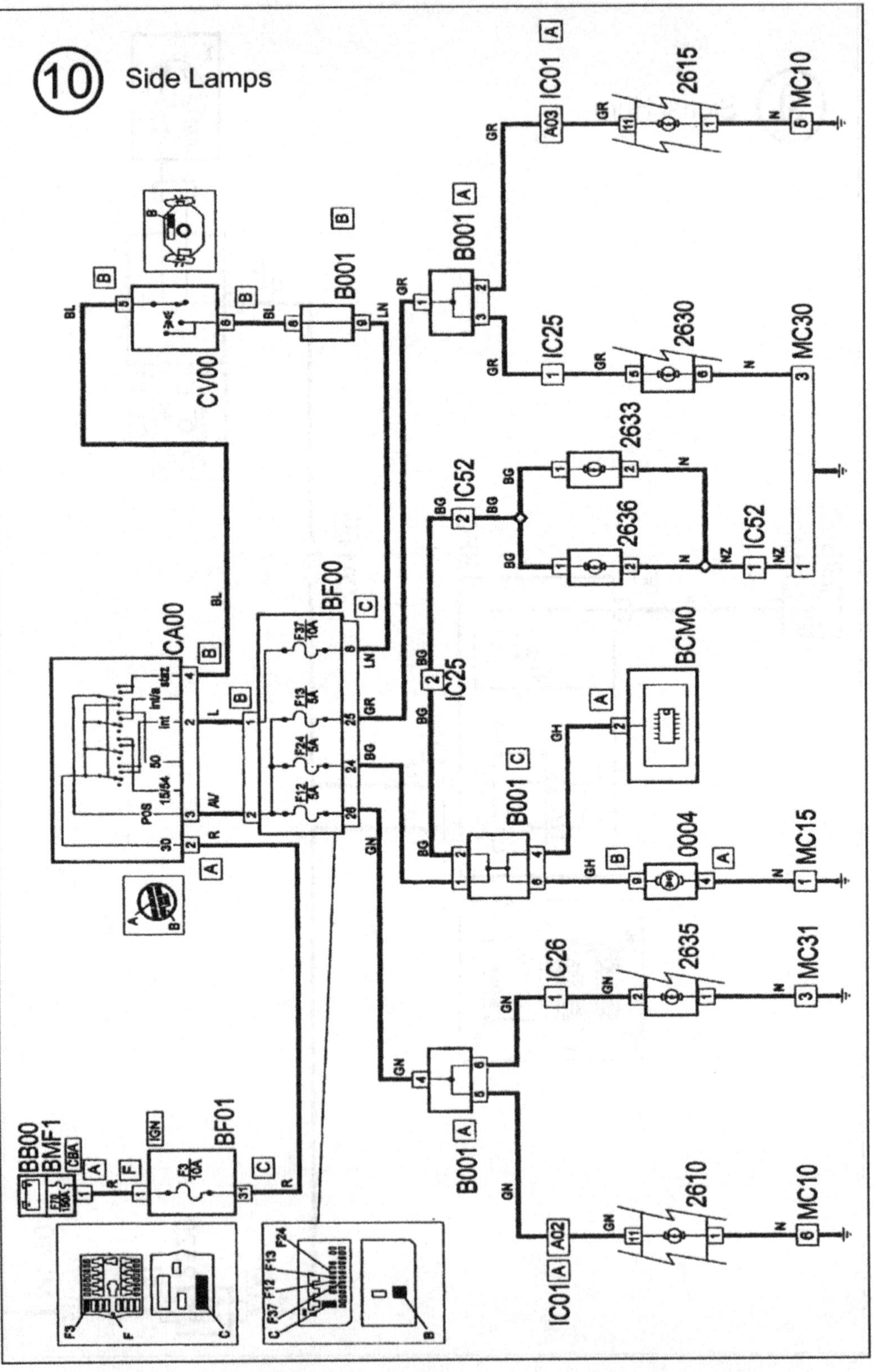

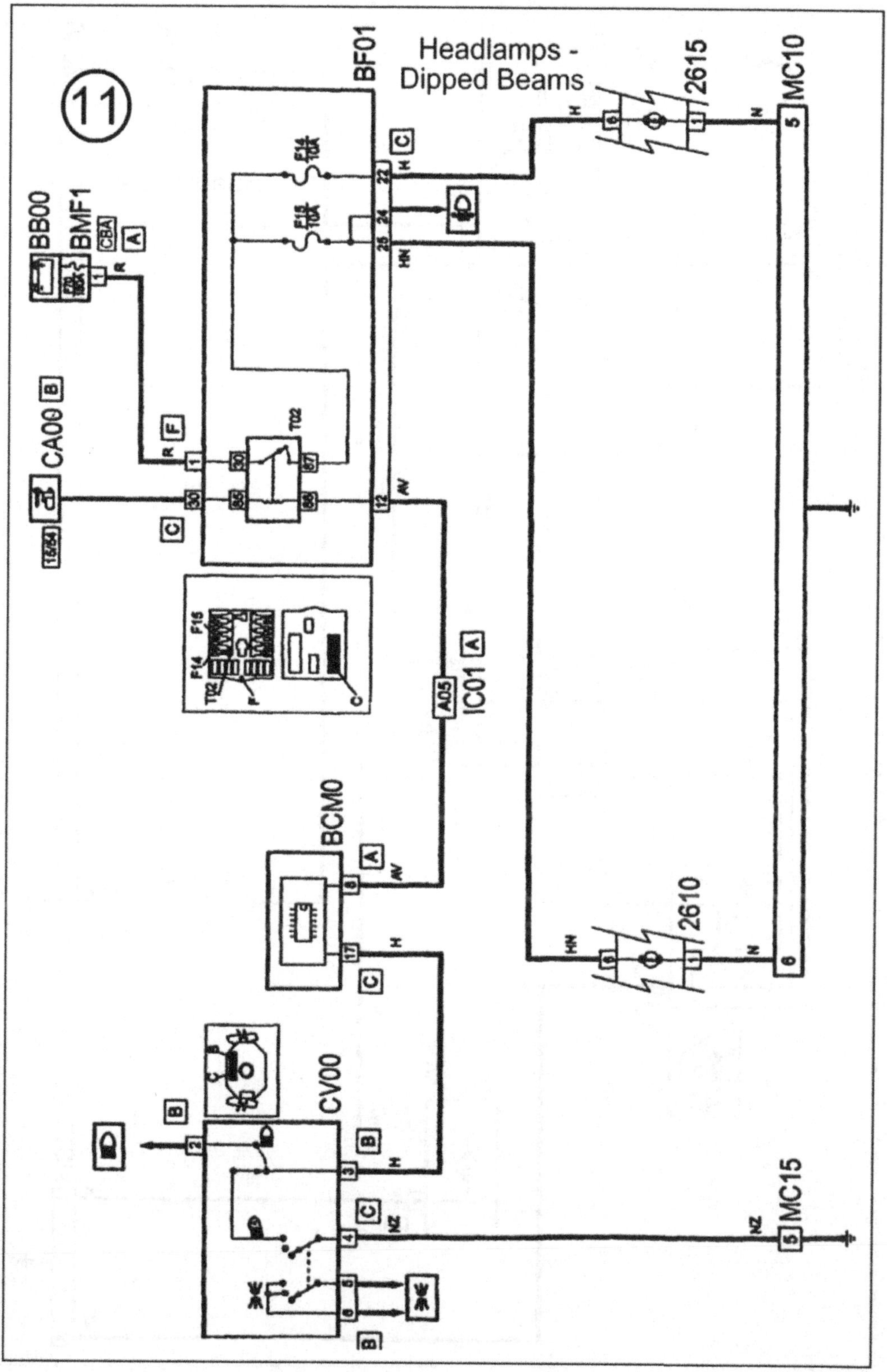

11
Headlamps - Dipped Beams
BF01
2615
MC10
BB00
BMF1
CBA
A
CA00
B
F
C
15/54
T02
F14 10A
F15 10A
IC01
A05
BCM0
CV00
2610
MC15

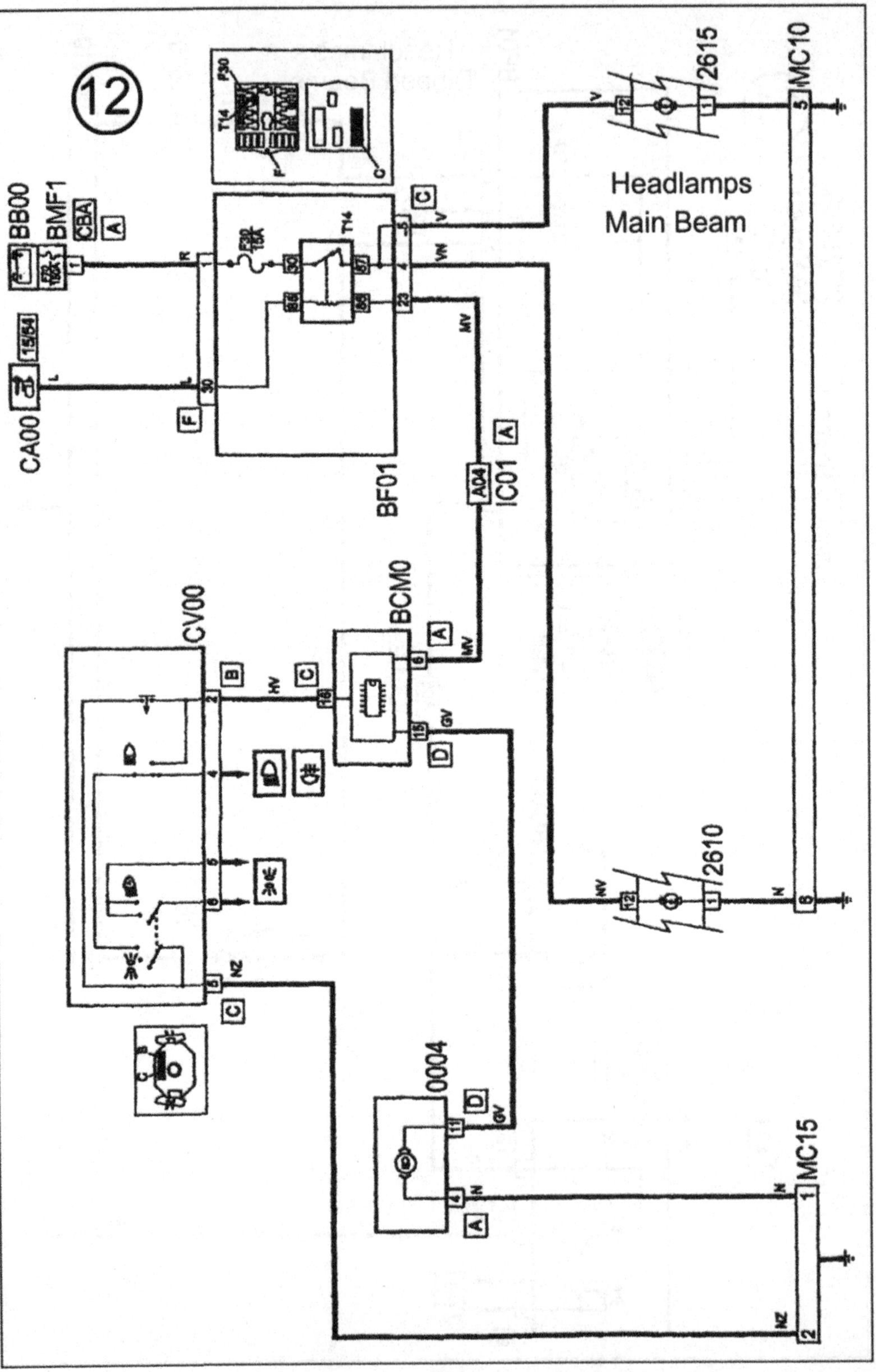
12
Headlamps
Main Beam
CA00
BB00
BMF1
CBA
15/54
BF01
A04
IC01
BCM0
CV00
0004
2615
2610
MC10
MC15

13 Instrument Panel Illumination

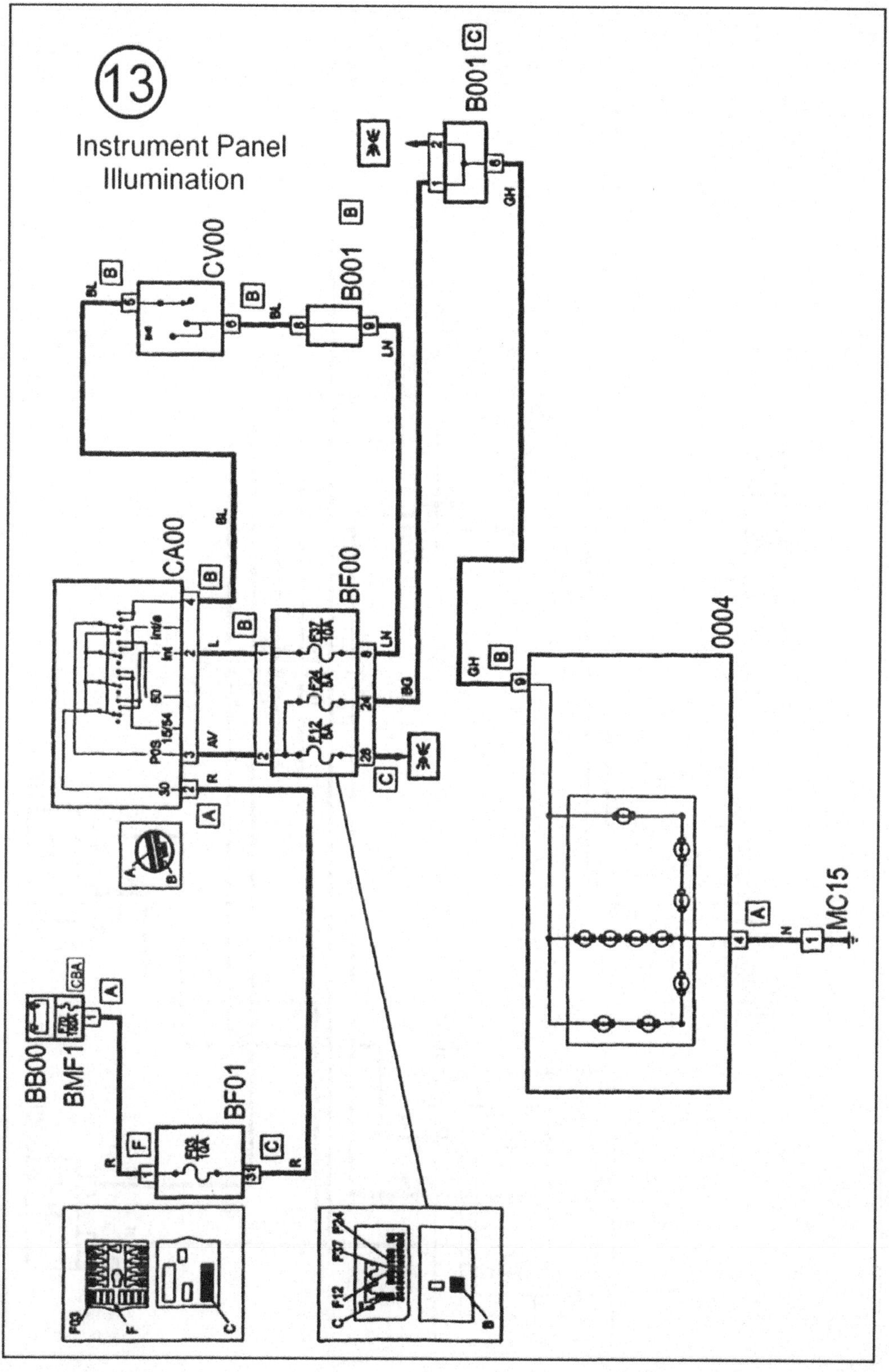

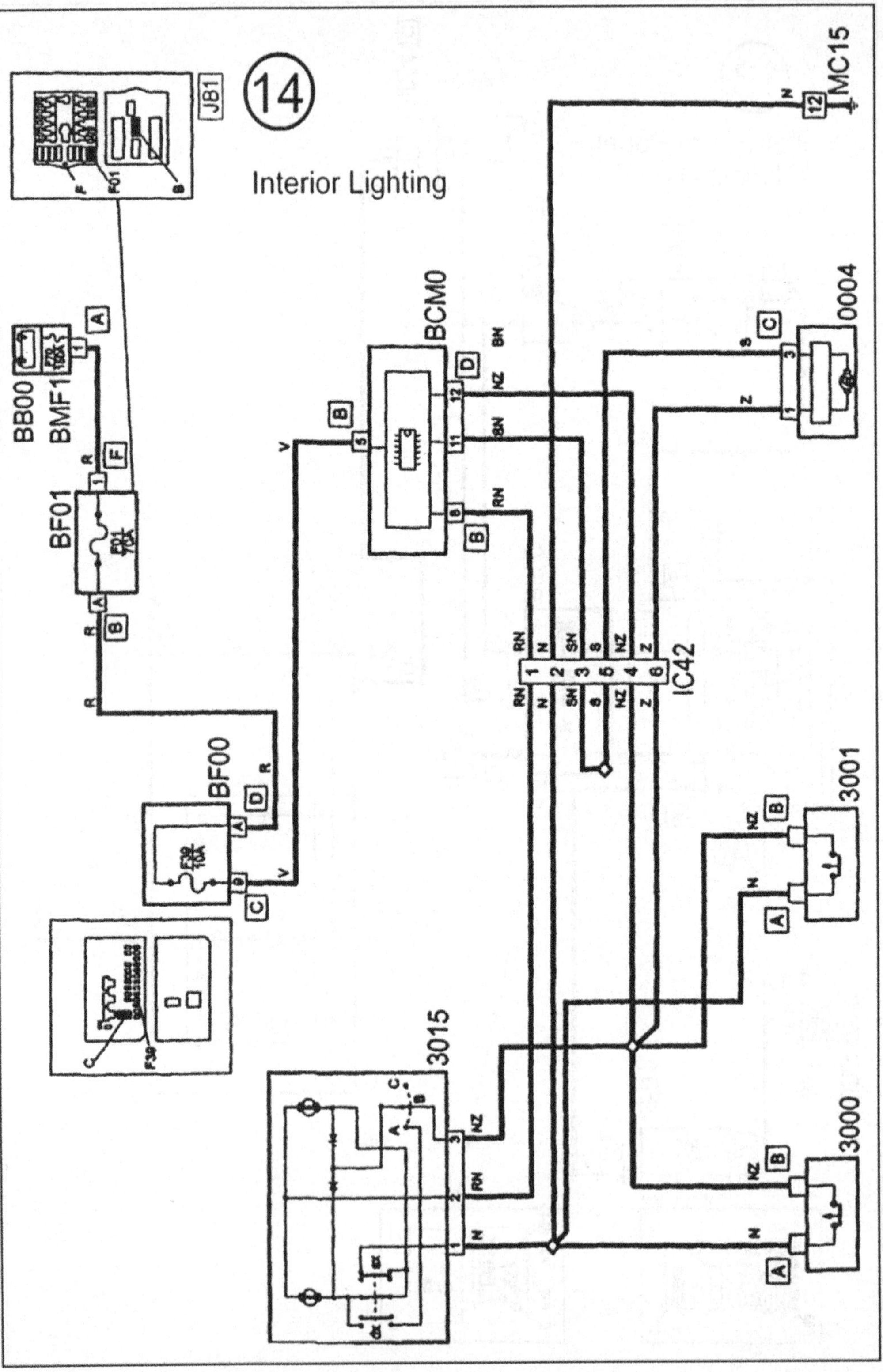
14
Interior Lighting
JB1
BB00
BMF1
BF01
BF00
BCM0
IC42
MC15
0004
3001
3000
3015

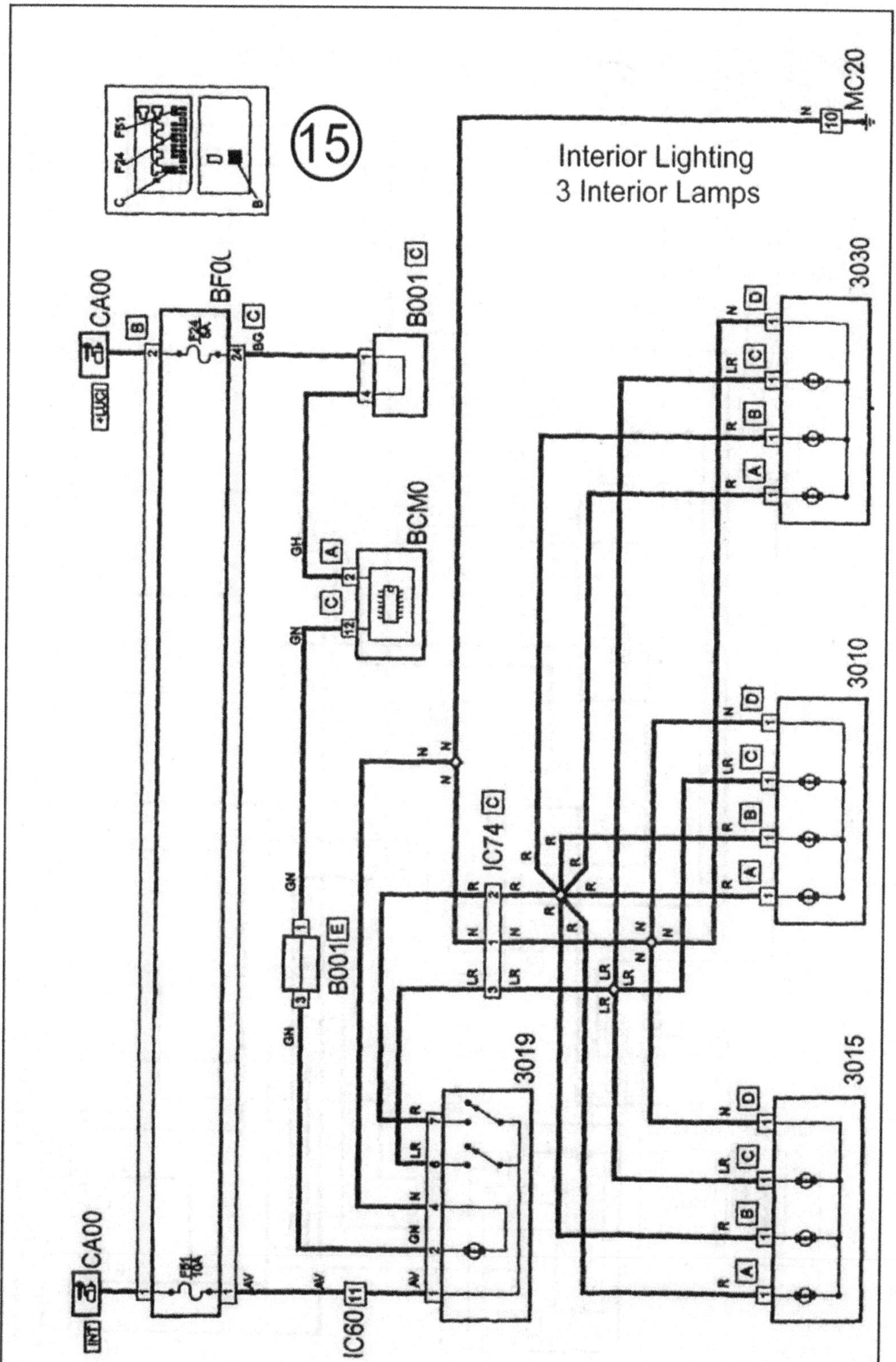
Interior Lighting
3 Interior Lamps
15
CA00
BF0L
B001
BCM0
IC74
IC60
MC20
3019
3015
3010
3030

Interior Lighting
4 Interior Lamps

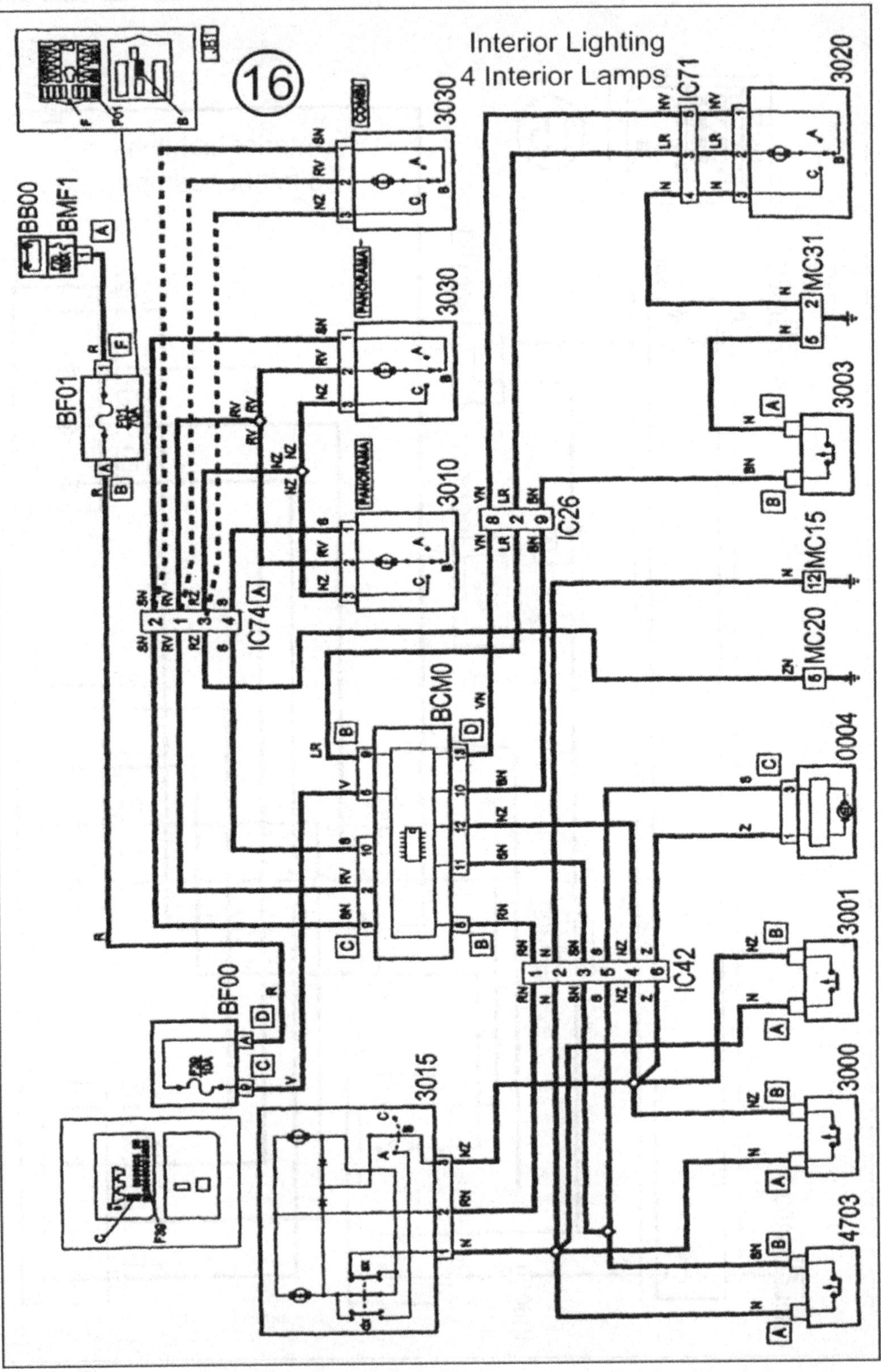

17 Front Screen Washer

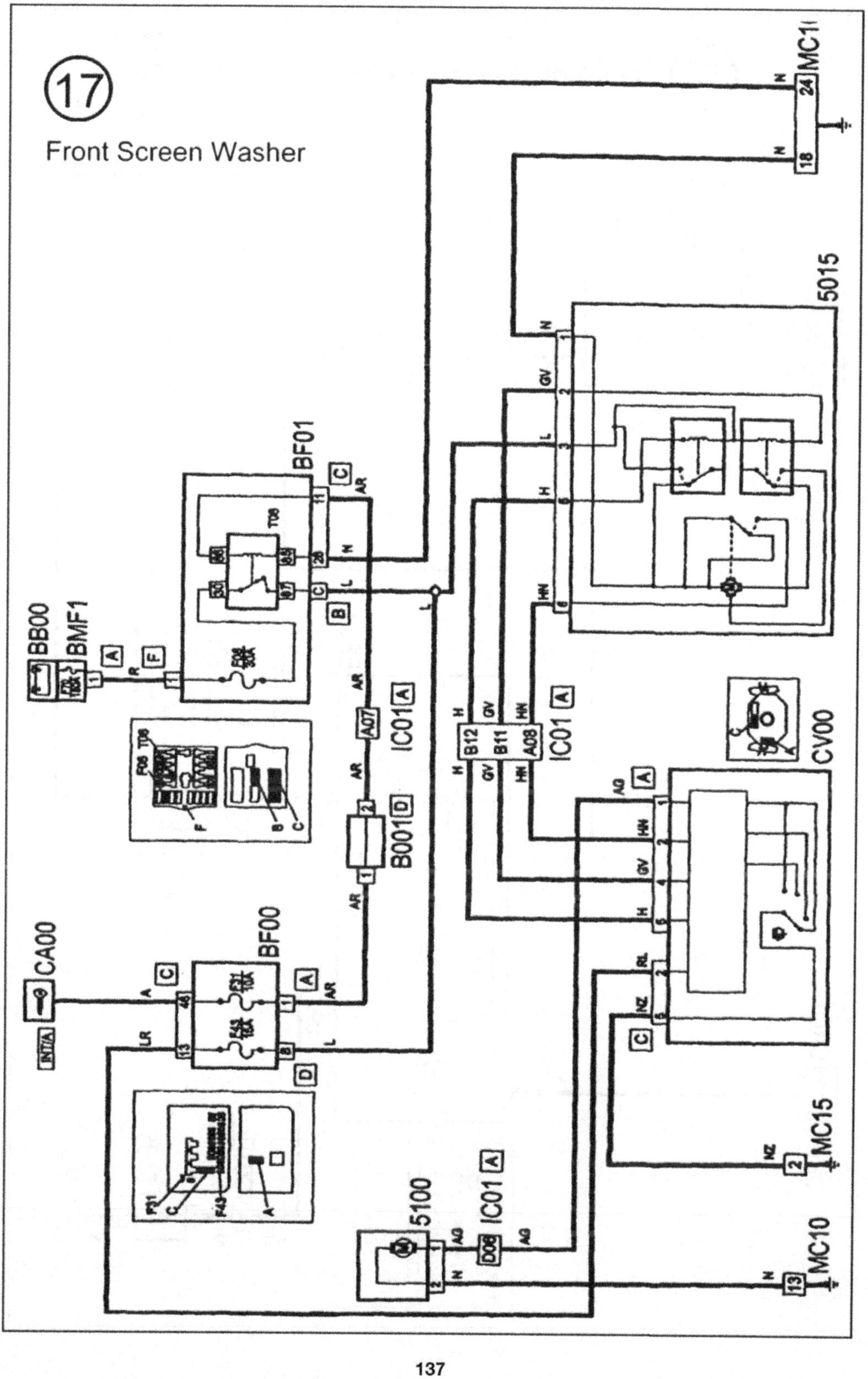

(18) Headlamp Washer System

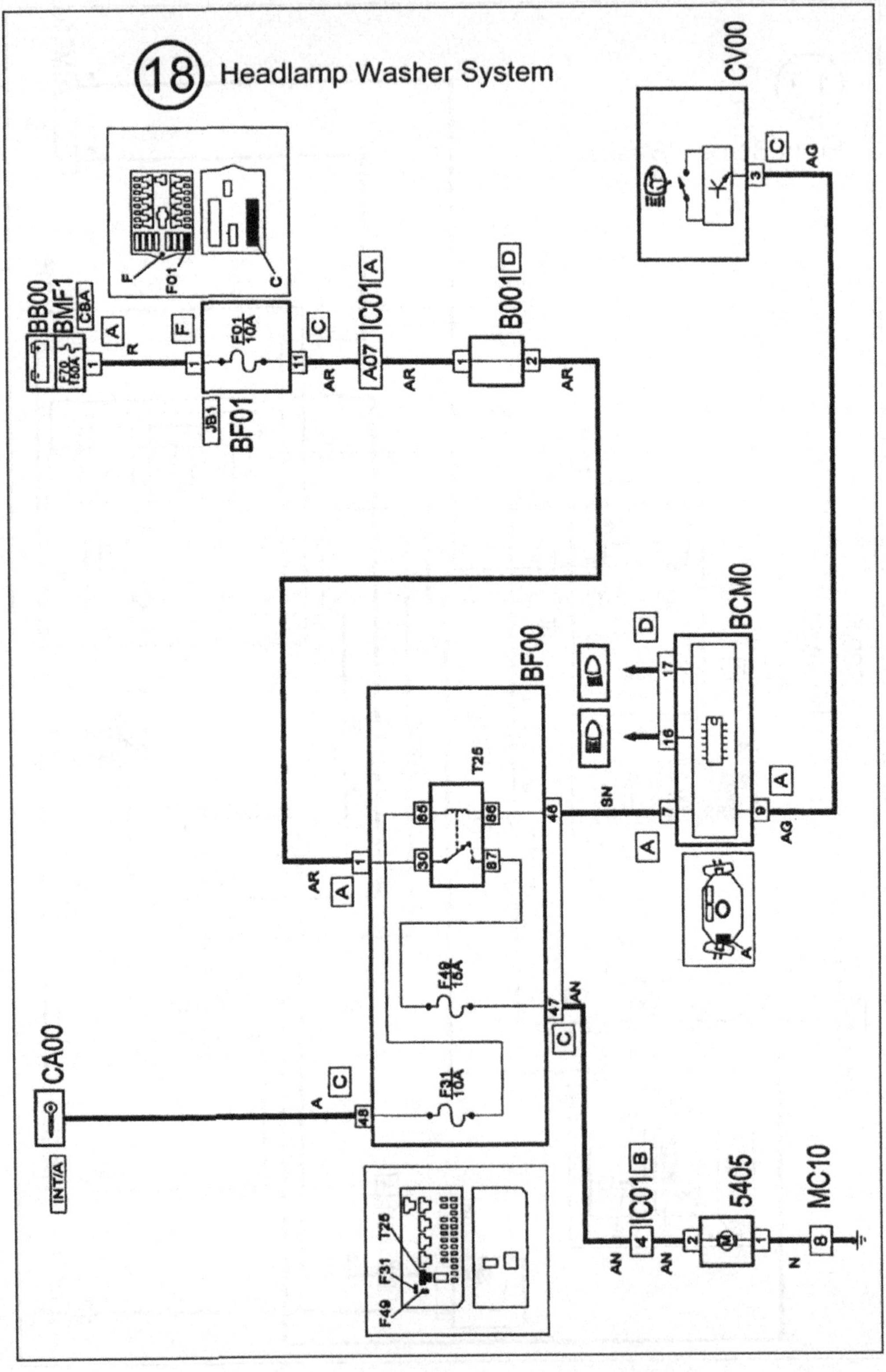

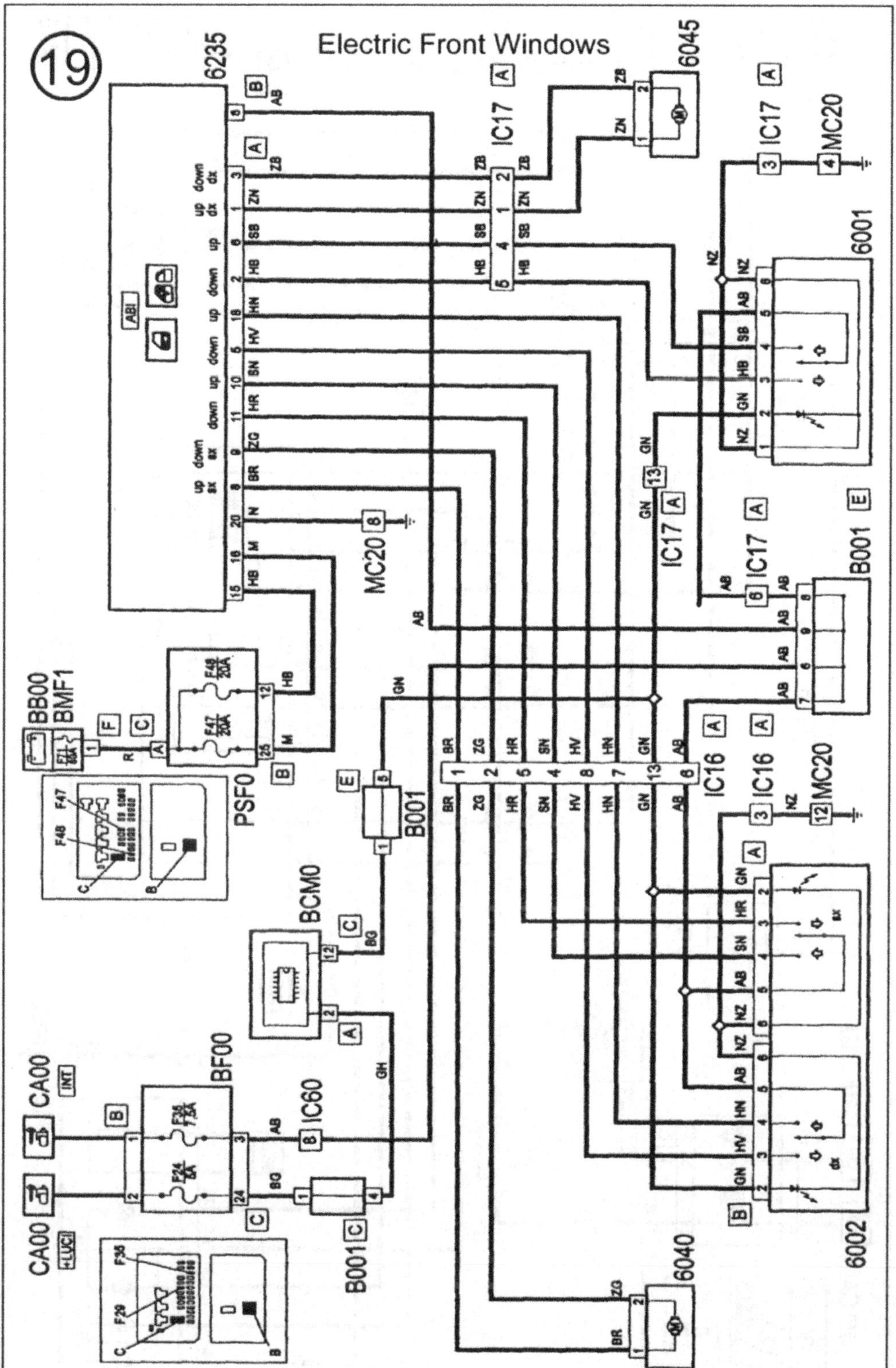
19
Electric Front Windows
6235
ABI
6045
6001
6002
6040
IC17
IC16
IC60
MC20
B001
BB00
BMF1
PSF0
BCM0
BF00
CA00
+LUCI
INT
F47 20A
F48 20A
F35 7.5A
F24 5A

20 Electically controlled Mirrors

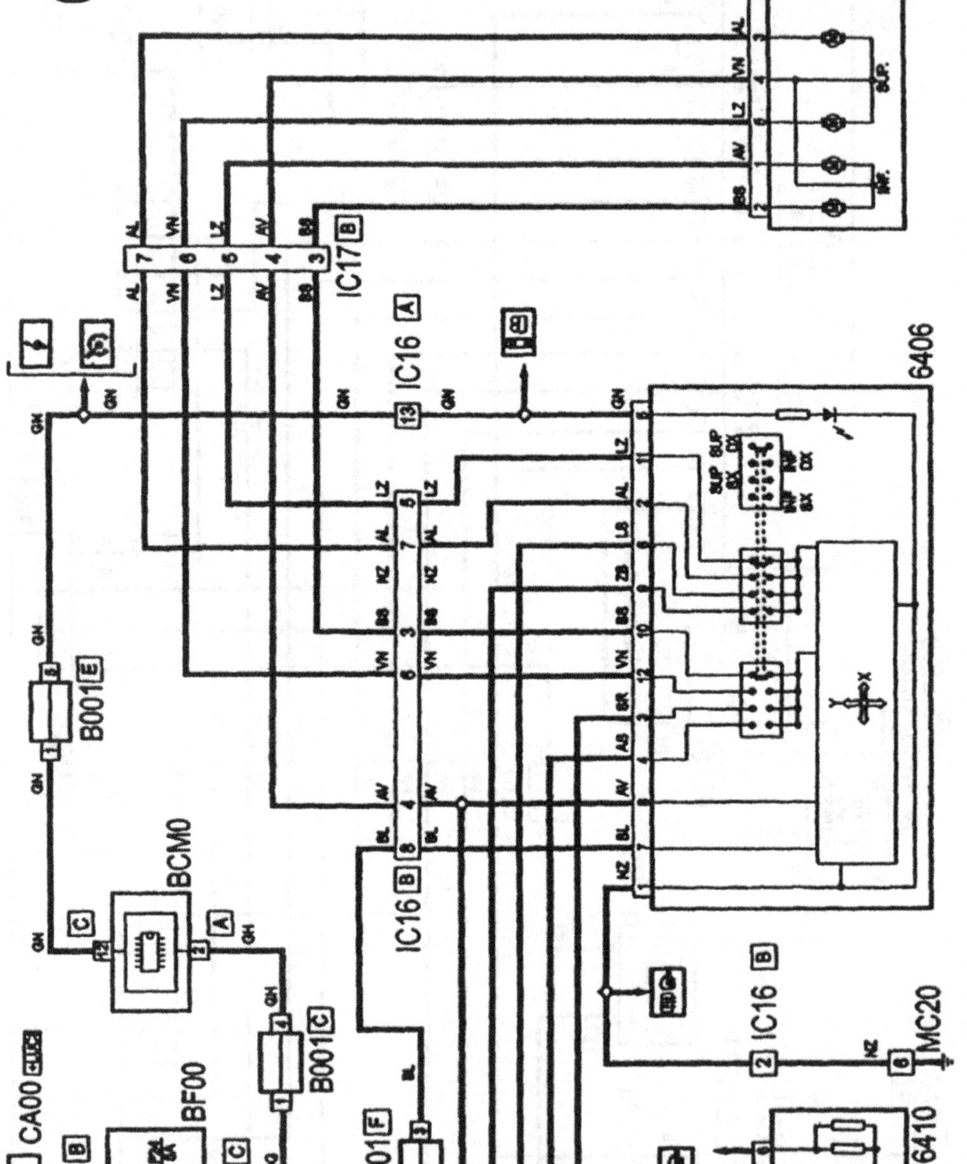

21 Anti-lock Brakes

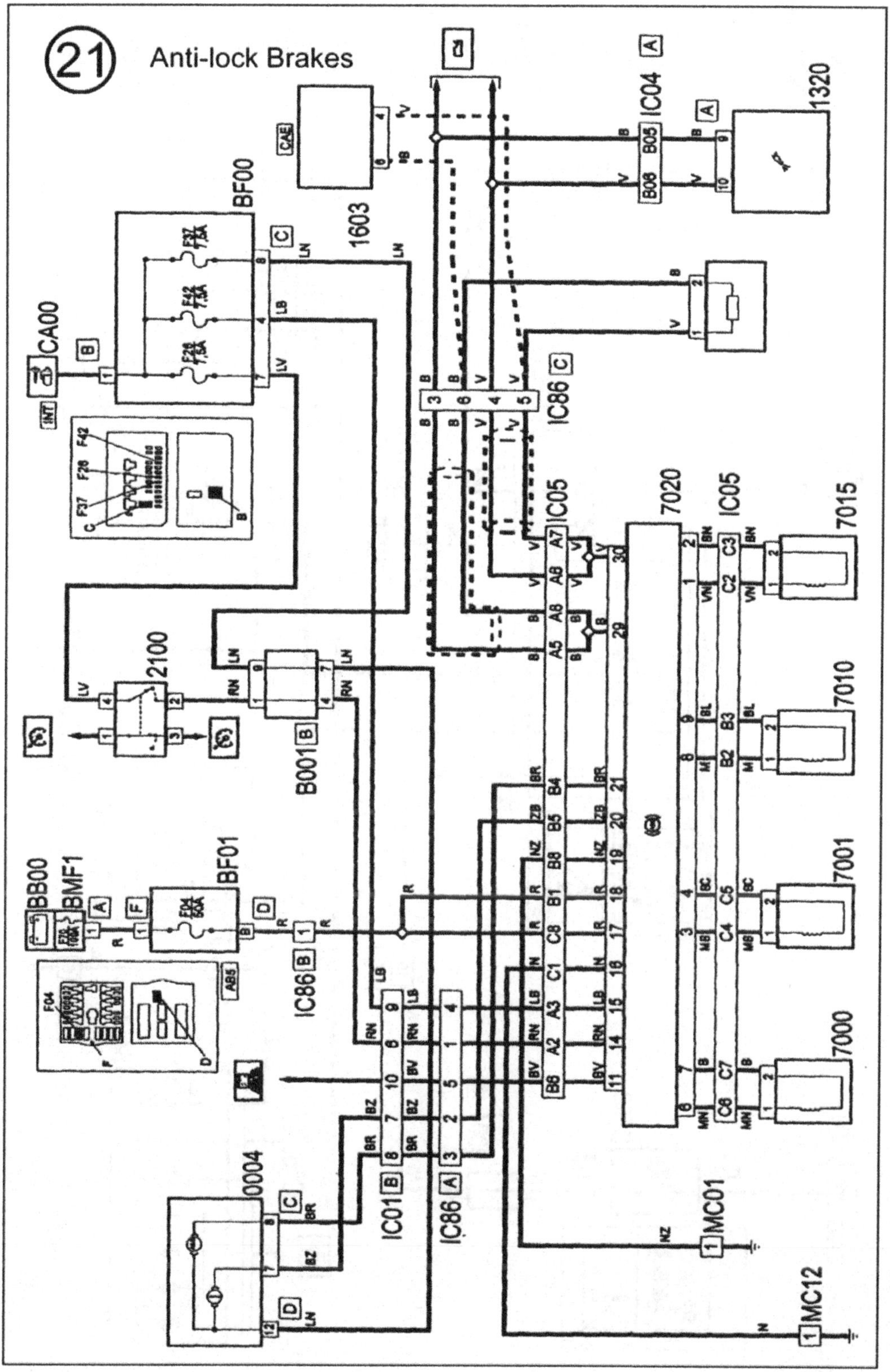

22

Heated rear Screen
Heated Mirrors

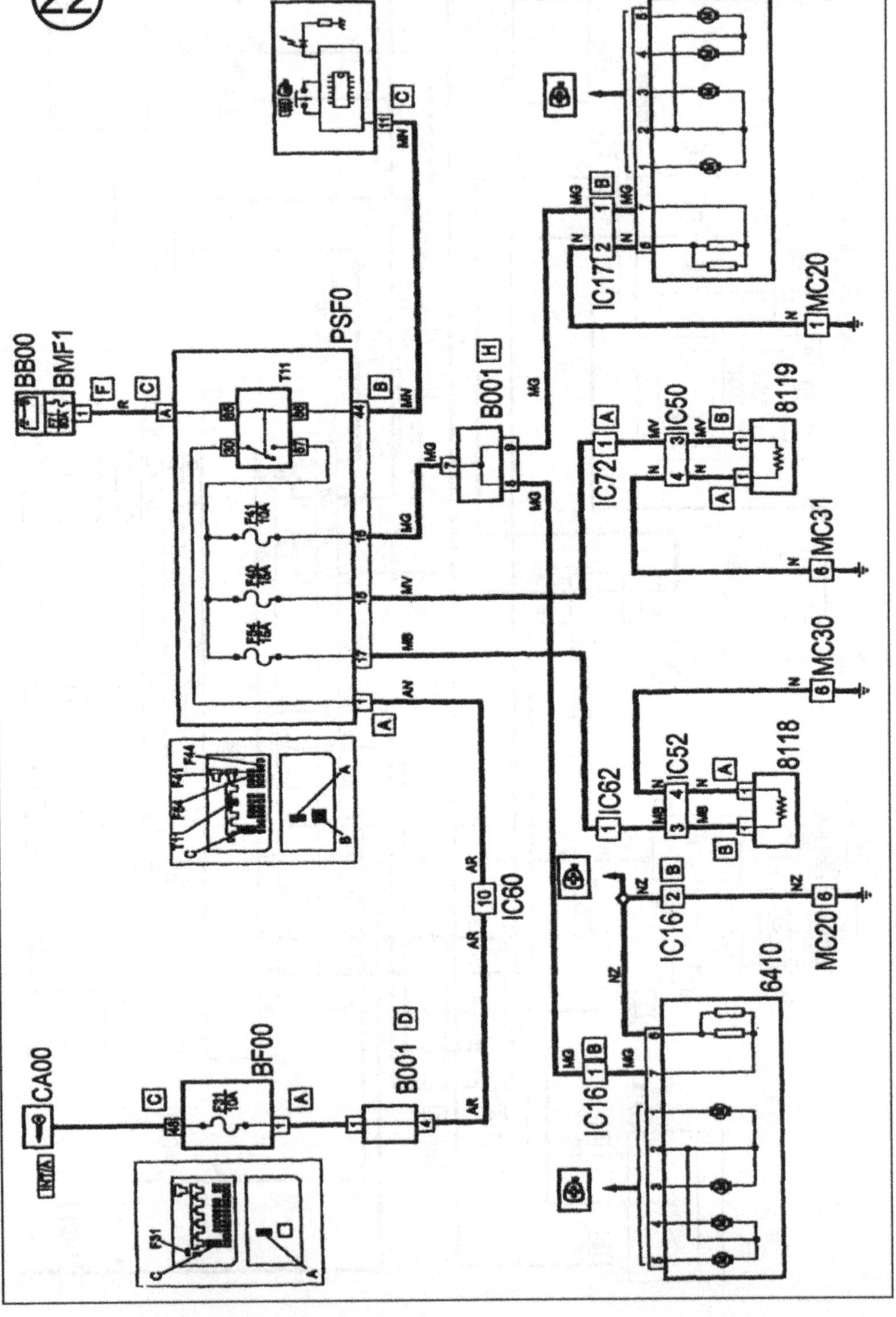

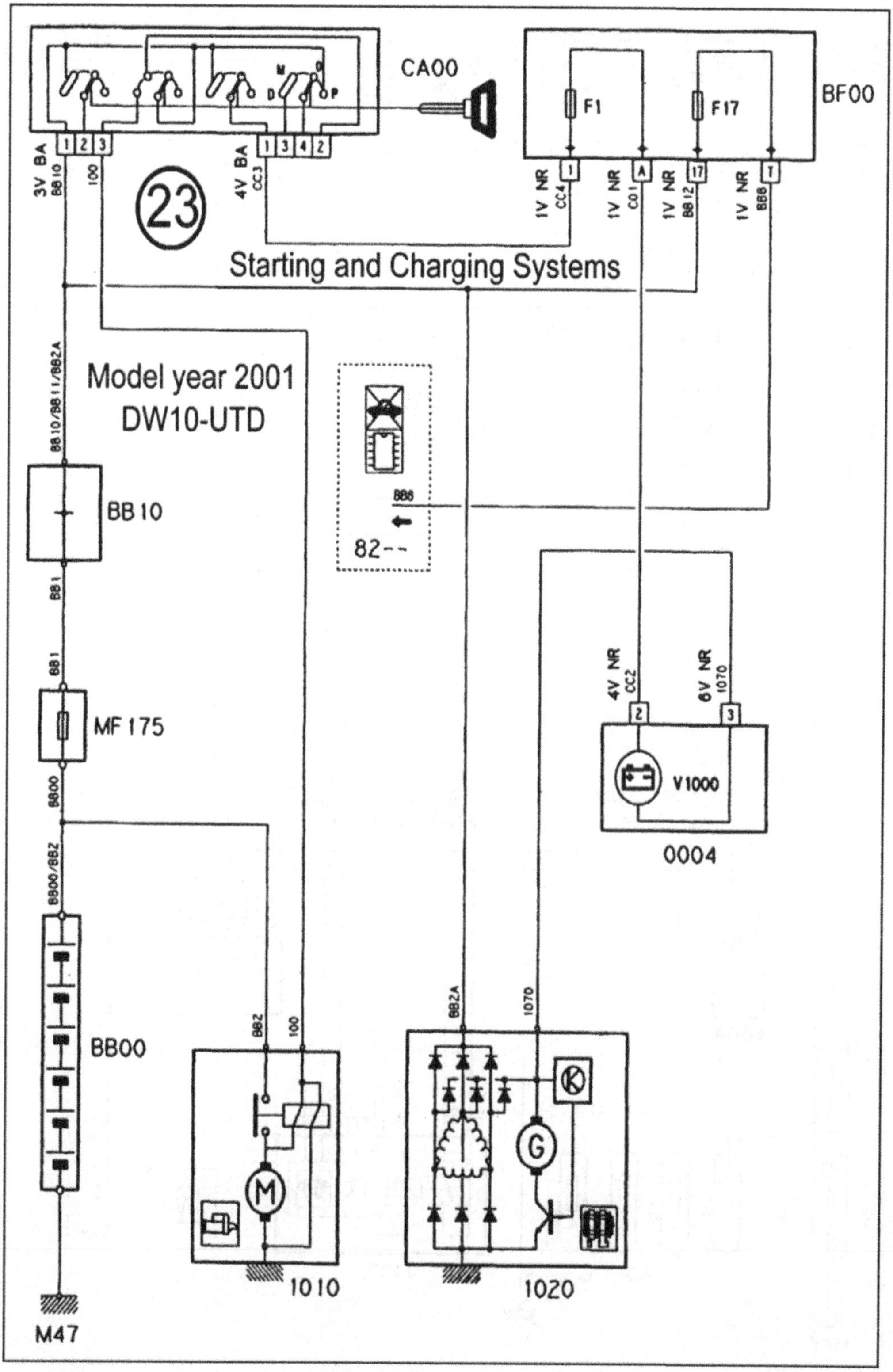
CA00
BF00
F1
F17
23
Starting and Charging Systems
Model year 2001
DW10-UTD
BB10
MF 175
BB00
M47
1010
1020
V1000
0004
82--
3V BA
BB10
100
4V BA
CC3
1V NR
CC4
C01
BB12
BB6
BB10/BB11/BB2A
BB1
BB00
BB00/BB2
BB2
BB2A
1070
4V NR
CC2
6V NR

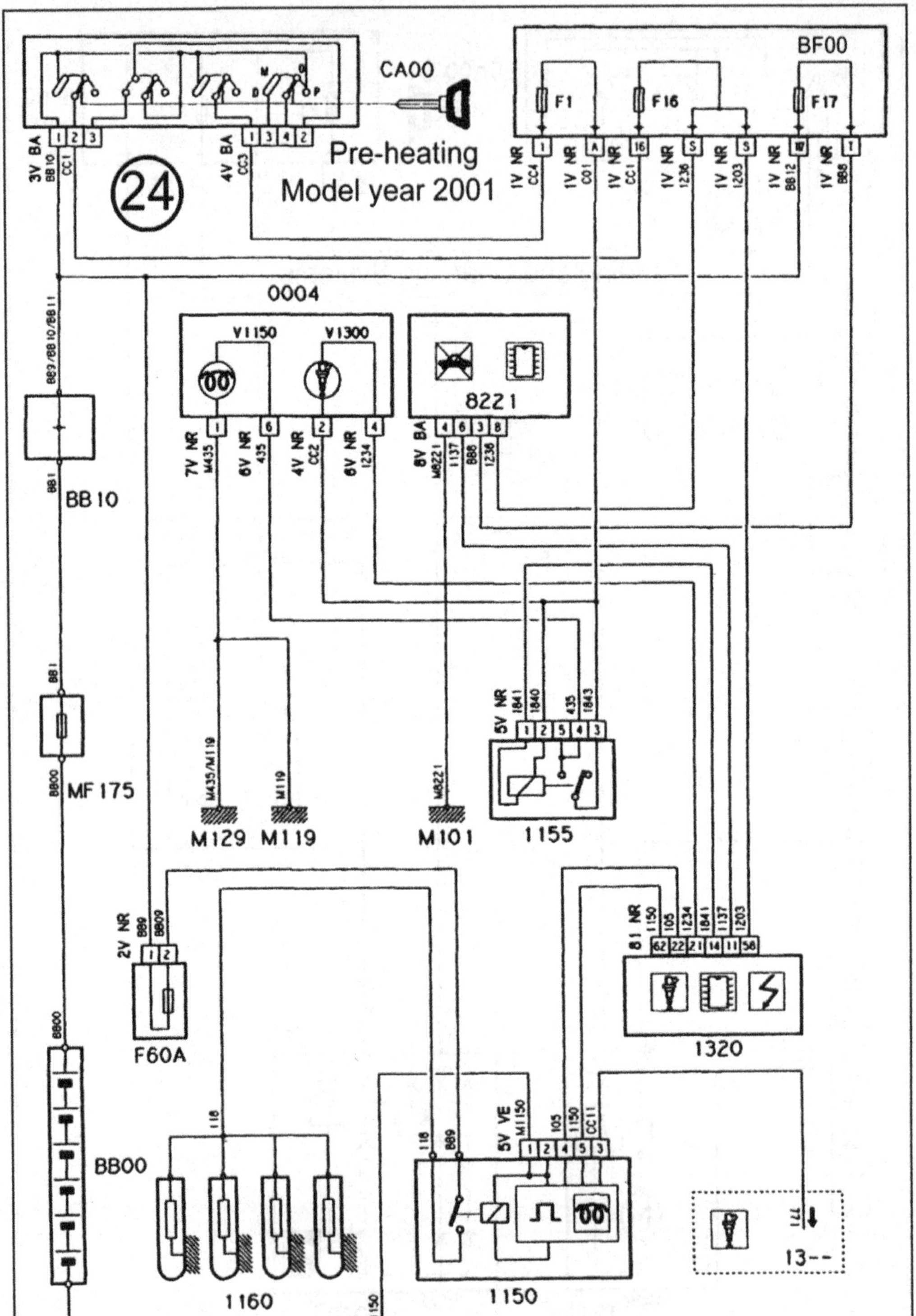
24
Pre-heating
Model year 2001
CA00
BF00
F1
F16
F17
0004
V1150
V1300
8221
BB10
MF 175
BB00
F60A
M129
M119
M101
M47
M7
1155
1320
1150
1160
13--

25a

Injection-Ignition
Model year 2001

cont. on 25b

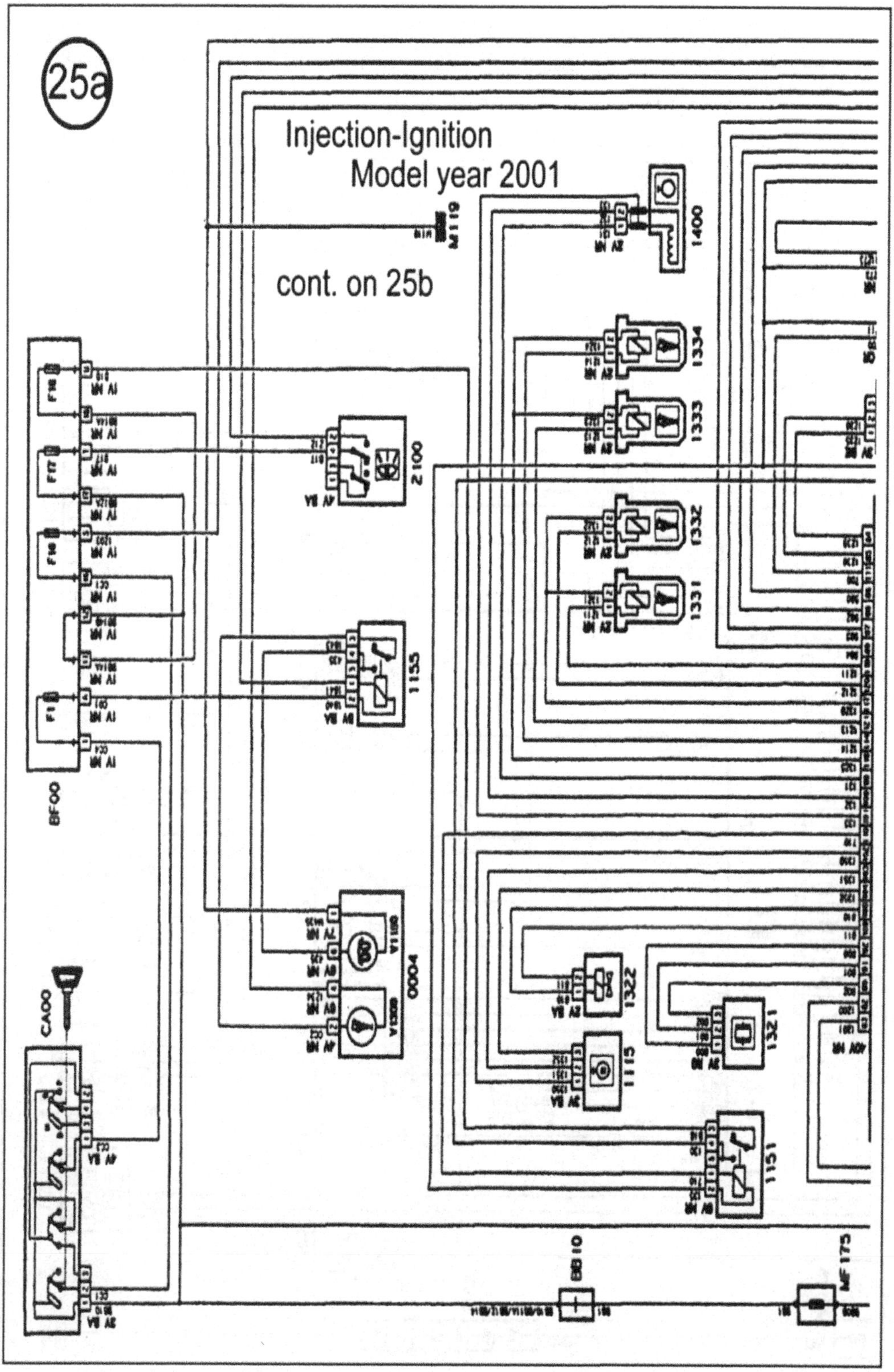

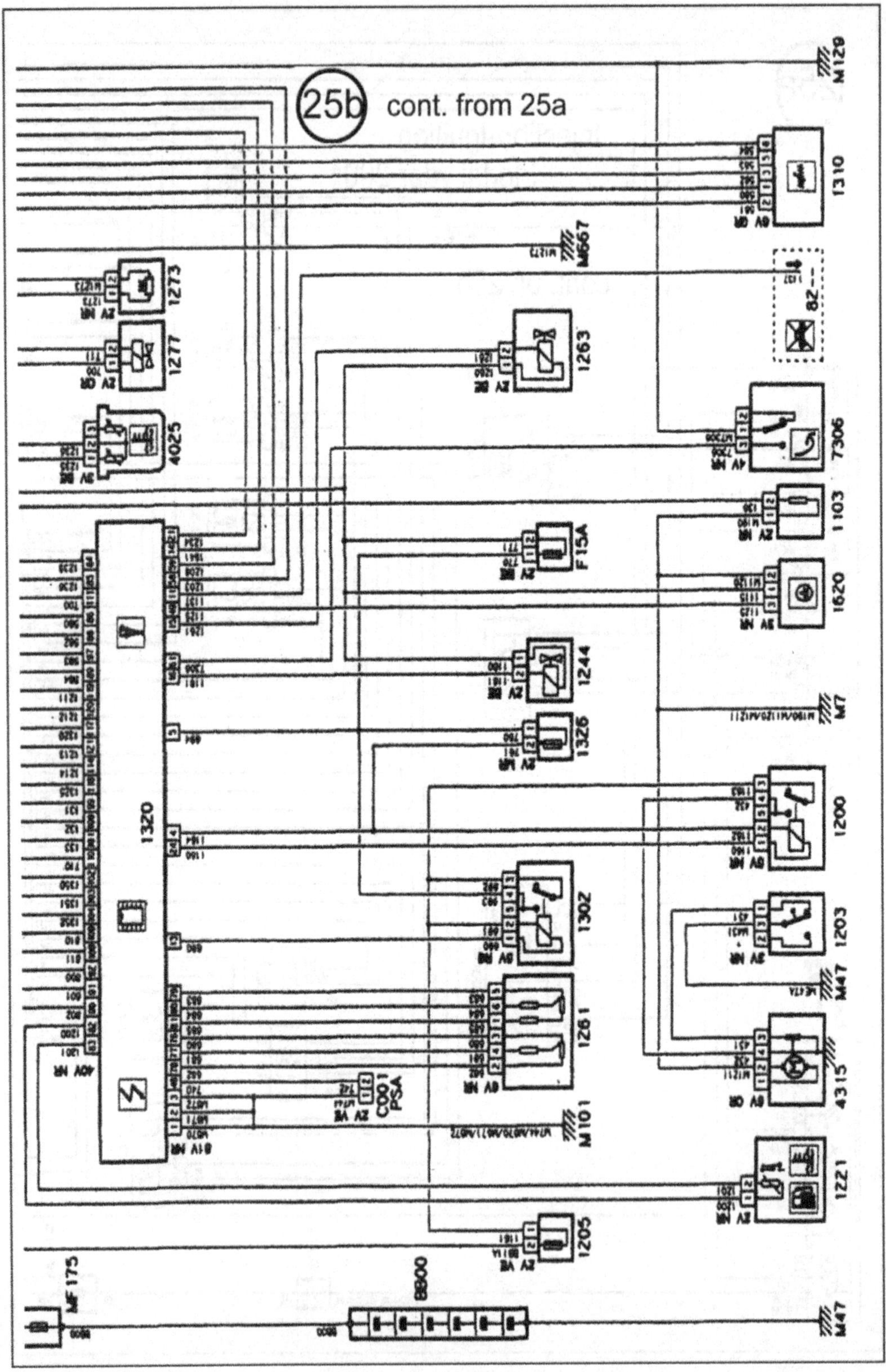
25b
cont. from 25a
M129
1310
M667
1273
1277
4025
1263
82
7306
1103
F15A
1620
1244
1326
M7
1320
1200
1302
1203
M47
1261
4315
C001
PSA
M101
1221
1205
BB00
MF175
M47

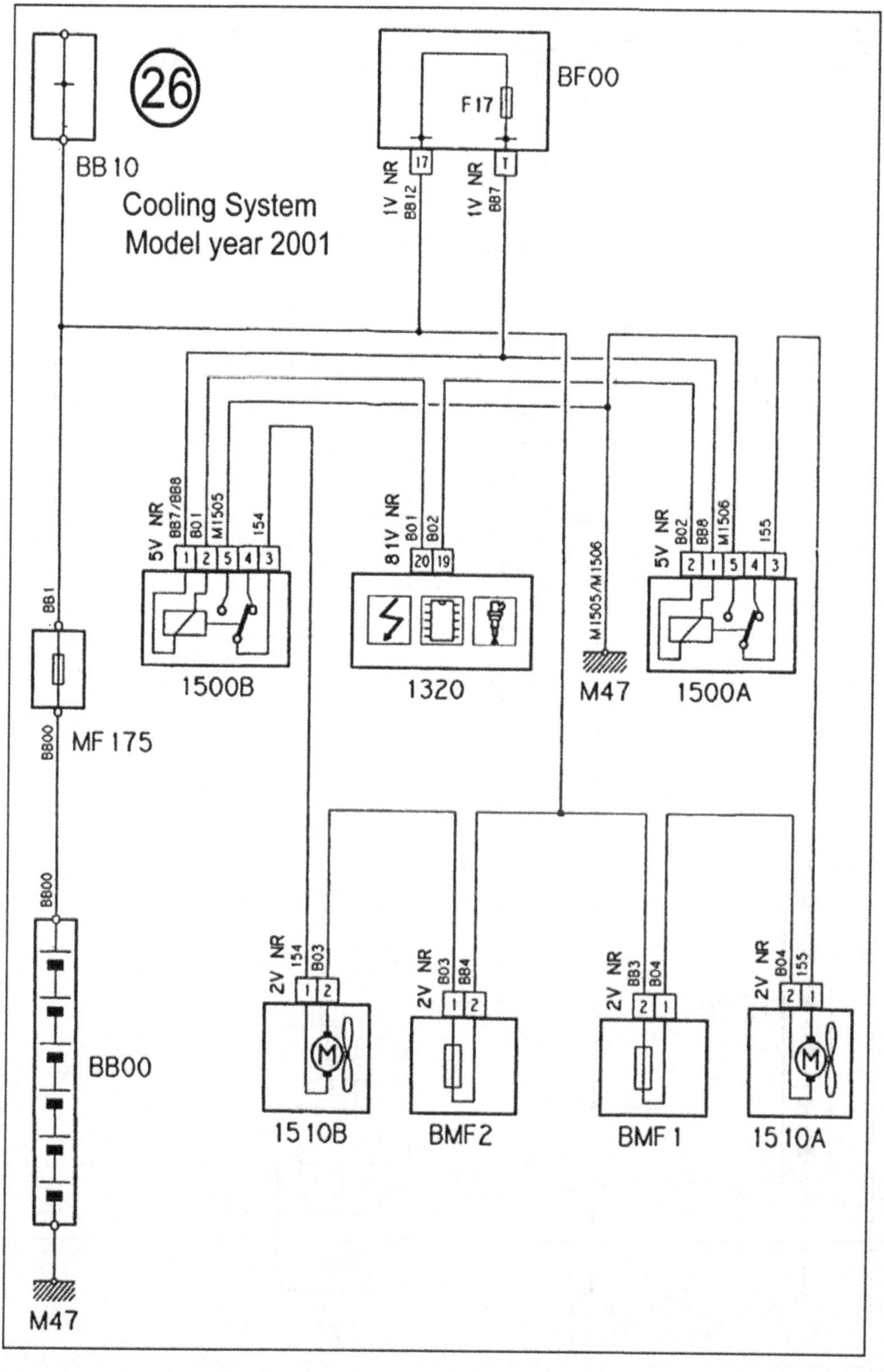

26
Cooling System
Model year 2001
BB10
BF00
F17
1V NR
BB12
BB7
17
1
5V NR
BB7/BB8
B01
M1505
154
1500B
81V NR
B01
B02
20
19
1320
M1505/M1506
M47
5V NR
B02
BB8
M1506
155
1500A
BB1
MF 175
BB00
2V NR
154
B03
1510B
2V NR
B03
BB4
BMF2
2V NR
BB3
B04
BMF 1
2V NR
B04
155
1510A
BB00
M47

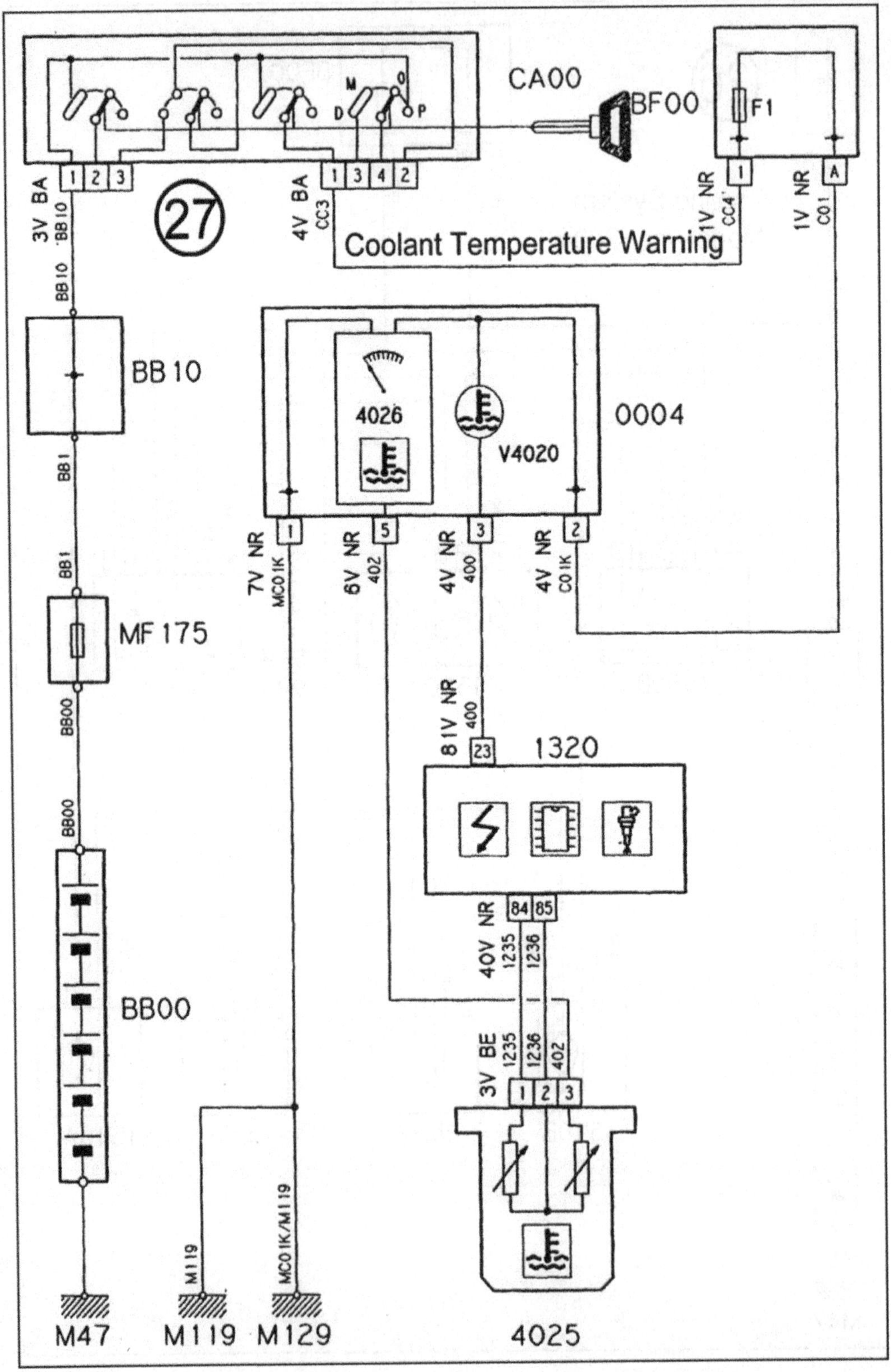
CA00
BF00
F1
27
Coolant Temperature Warning
3V BA
BB10
4V BA
CC3
1V NR
CC4
1V NR
C01
BB10
0004
4026
V4020
BB1
MF 175
7V NR
MC01K
6V NR
402
4V NR
400
4V NR
C01K
BB00
81V NR
400
1320
40V NR
1235
1236
BB00
3V BE
1235
1236
402
M119
MC01K/M119
M47
M119
M129
4025

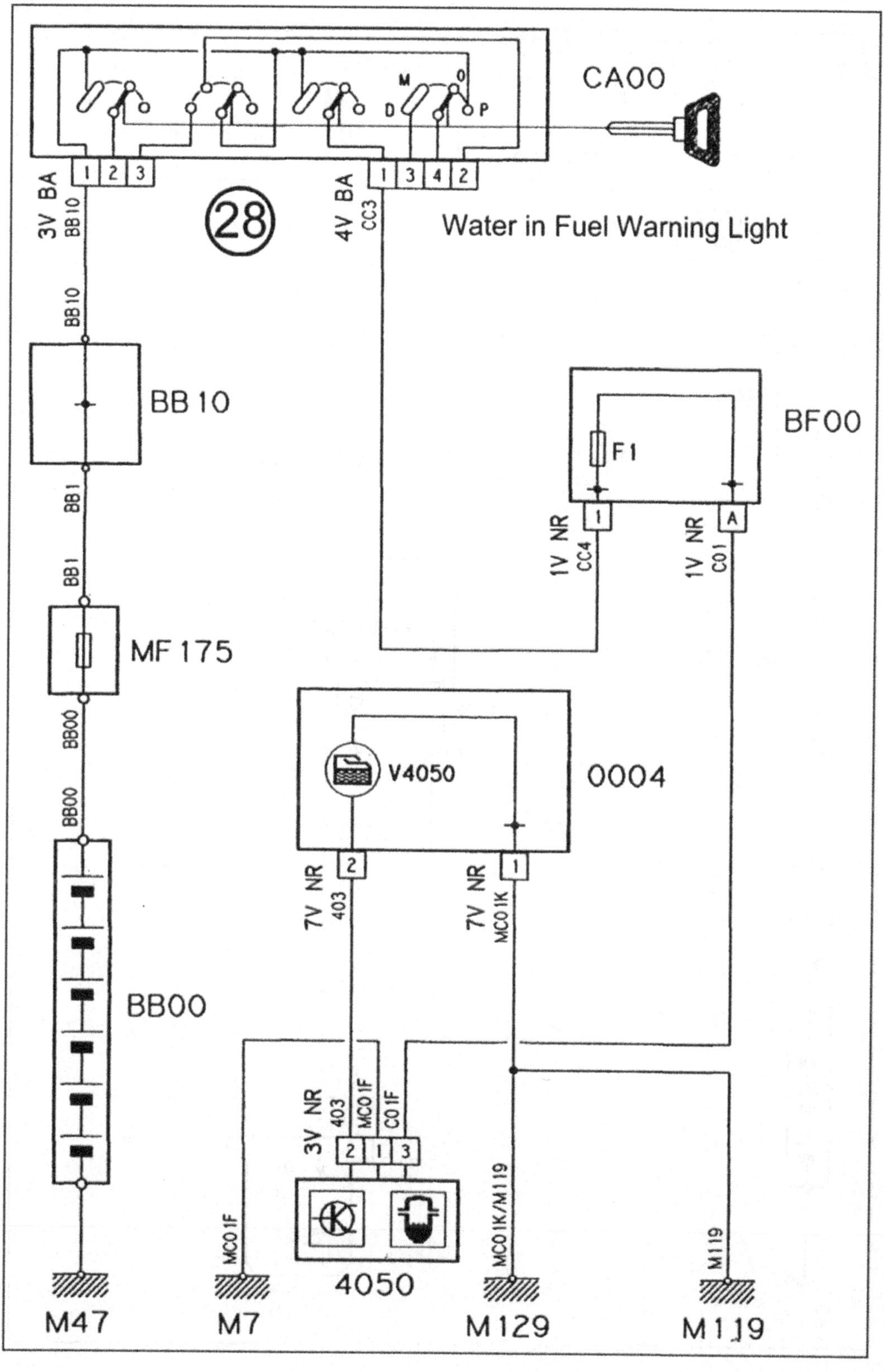
CA00
28
Water in Fuel Warning Light
3V BA
BB10
4V BA
CC3
M
O
D
P
BB10
BB1
MF 175
BB00
BF00
F1
1V NR
CC4
1V NR
C01
V4050
0004
7V NR
403
7V NR
MC01K
3V NR
403
MC01F
C01F
MC01F
4050
MC01K/M119
M119
M47
M7
M129
M119

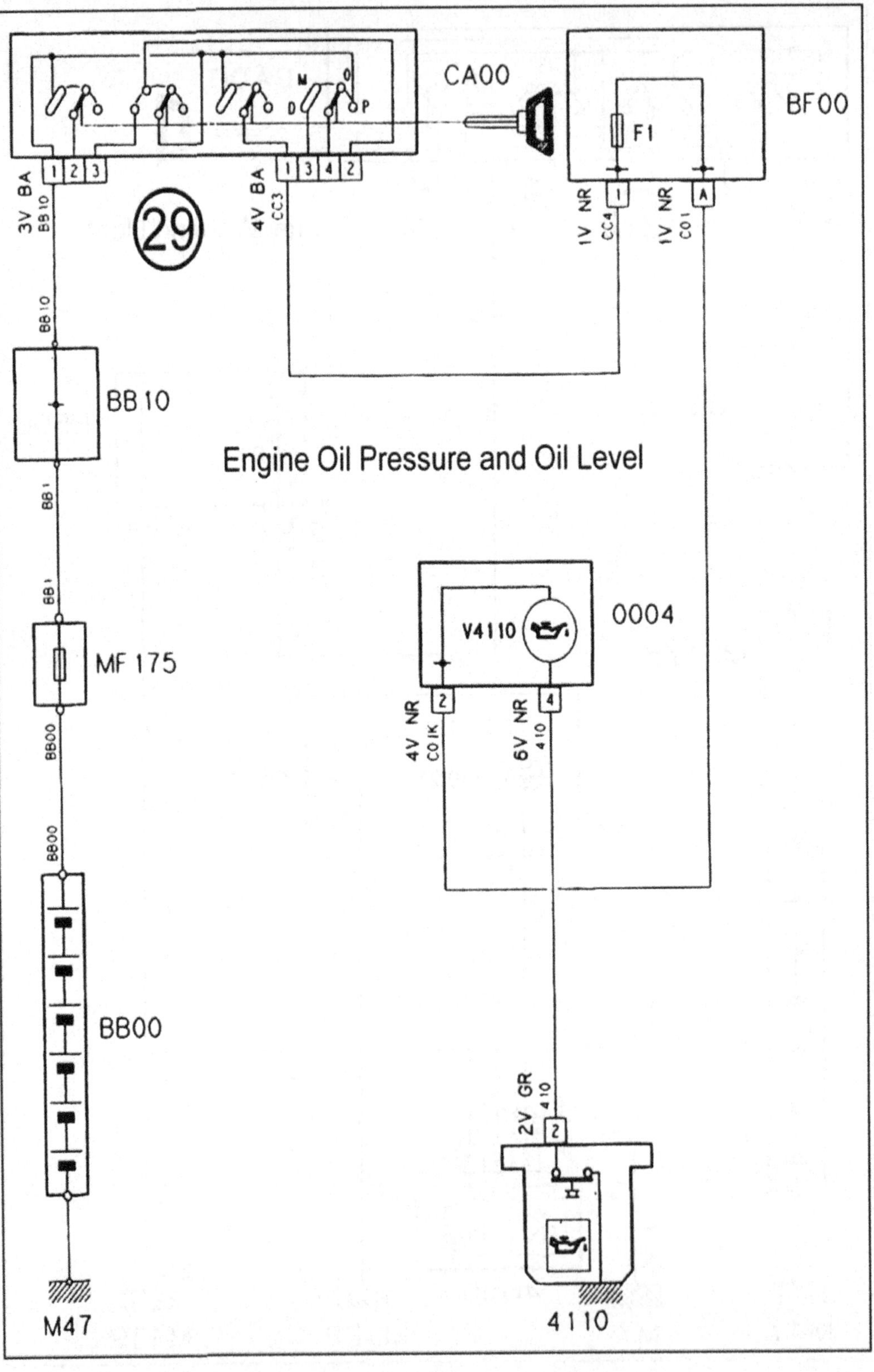

Engine Oil Pressure and Oil Level

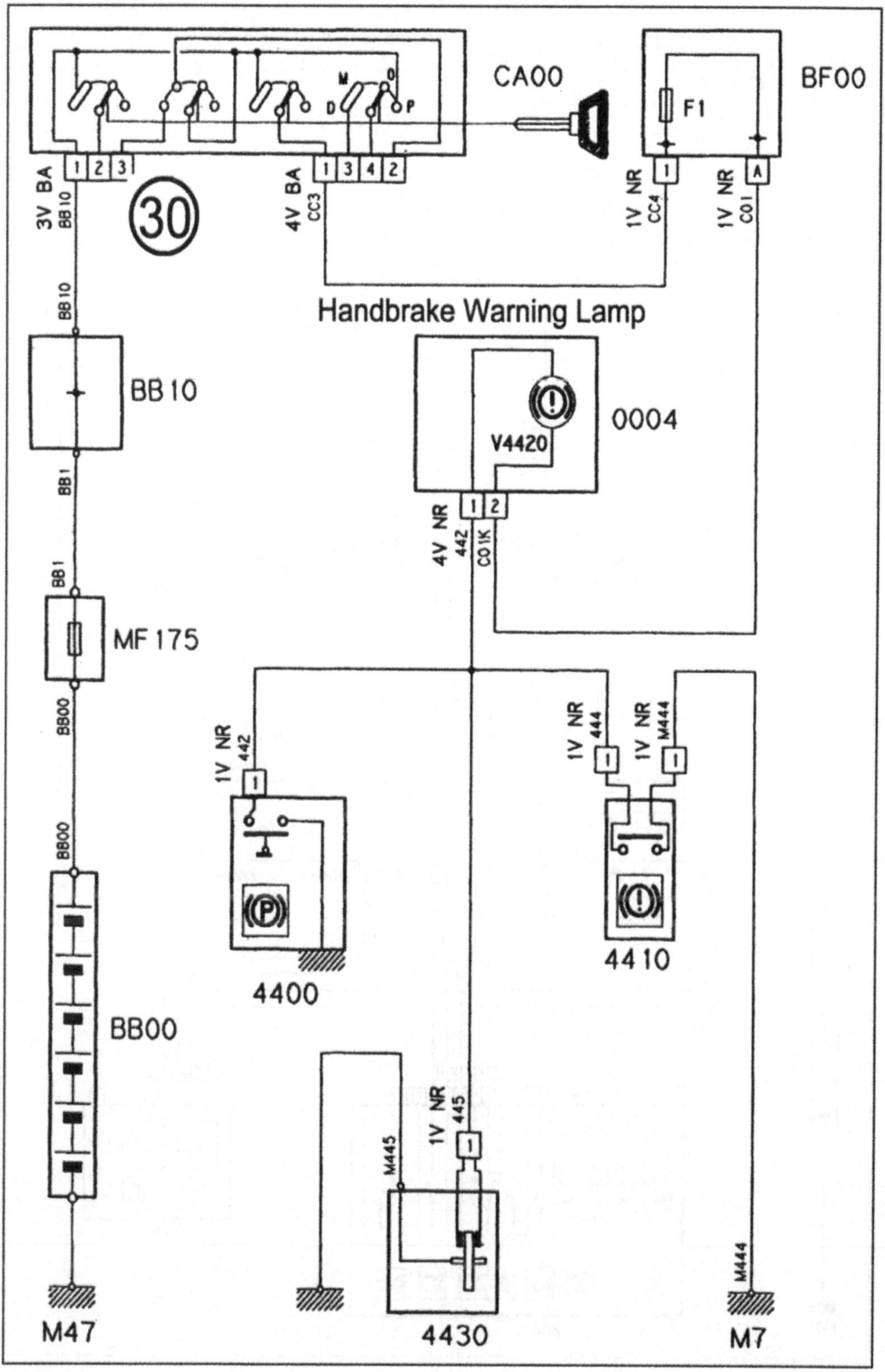
CA00
BF00
F1
30
Handbrake Warning Lamp
BB10
0004
V4420
MF175
BB00
4400
4410
4430
M47
M7
3V BA
4V BA
1V NR
4V NR
1V NR
442
444
M444
445
M445
CC3
CC4
C01
C01K
BB1

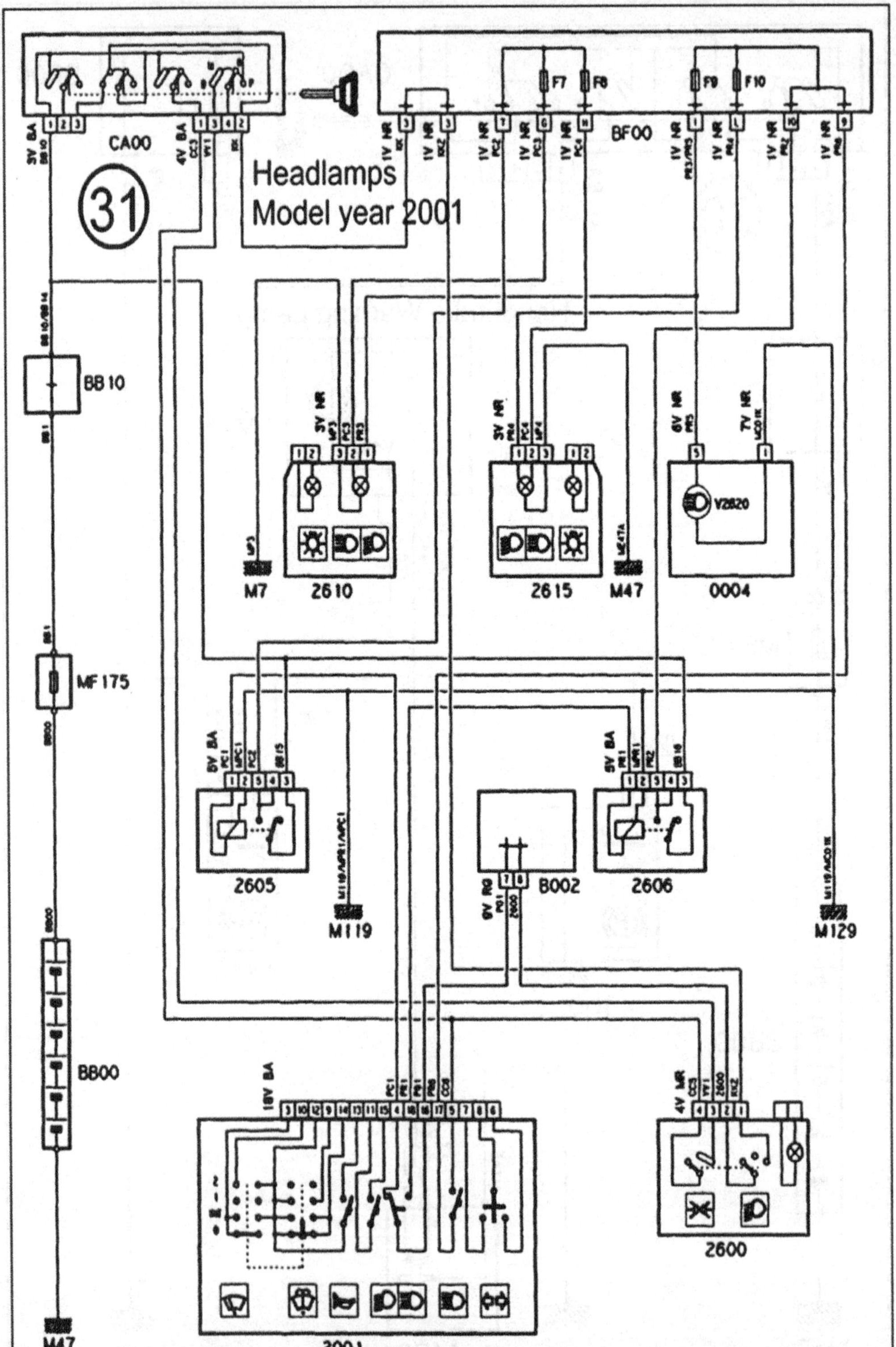

31
Headlamps
Model year 2001
CA00
BF00
F7
F8
F9
F10
BB10
MF 175
BB00
M47
M7
2610
2615
M47
0004
V2620
2605
2606
B002
M119
M129
2001
2600

FULL CONTENTS

UPCOMING TITLES FROM
THE POCKET MECHANIC

Audi A3 1.9 Litre Diesel from Model Year 2001-2004	9781907070037
Citroen Berlingo, Peugeot Partner Diesel Models 2000-2004	9781989780602
1.8 Litre, 1.9 Litre & 2.0 Diesel XUD7, XUD9, DW8, DW8B & DW10 HDi Engines (Bosch & Siemens Injection)	
Citroen Berlingo, Peugeot Partner Petrol Models 1996-2004	9781898760619
1.1, 1.4, 1.6 (16V) & 1.8 Litres	
Citroen Dispatch, Peugeot Expert, Fiat Scudo Diesel Models 1995-2001	9781898780755
1.9 Litre Diesel and Turbo diesel & 2.0 Litre HDi Diesel	
Citroen Relay, Peugeot Boxer & Fiat Ducato 2.0L HDI (JTD) Diesel from 2001	9781898780984
Citroen Relay, Peugeot Boxer & Fiat Ducato 2.5 & 2.8HDi Diesel 1994-2001	9781898780946
Citroen Relay, Peugeot Boxer & Fiat Ducato Diesel from 2001	9781898780946
2.0 DW10 HDI & 2.2 DW12 HDI Diesel	
Citroen Synergie, Peugeot 806, Fiat Ulysse & Lancia Z 1995-2001	9781898780809
1.9 Diesel & Turbo diesel & 2.0 DW10 HDi (8 & 16 Valve) & 2.1 XUD11	
Land Rover Freelander 1.8 Petrol & 2.0 Litre Turbo Diesel from 1997	9781907070020
Mazda E2000/E2200 Van & Bus Petrol & Diesel	9780907779865
2.0 Litre Petrol and 2.2 Litre Diesel FE and R2 Engine	
Mercedes A Class & Vaneo 1.7L CDI Diesel 1998-2004	9781898780472
A160 CDi, A170 CDi, A170L CDi	
Mercedes A Class & Vaneo 1.4, 1.6 & 1.9L Petrol 1998-2004	9781898780427
A140, A140L, A160, A160L, A190 & A190L	
Mercedes Sprinter & TL 2.3 & 2.9L Diesel up to 2000	9781898780892
208D, 210D, 212D, 308D, 310D, 408D, 410D, 412D 601 & 602 with Injection Pump	
Mitsubishi Colt 1.3, 1.6 & 1.8 Litre Engines from 1992	9781898780496
Mitsubishi Shogun, Pajero, Montero 4WD Models 1999-2003	9781898780823
3.5-litre GDI V6 petrol & 2.5, 2.8 & 3.2L Diesel Models	
Mitsubishi Space Runner & Space Wagon from 1992	9781898780366
1.8L 16V & 2.0 16V & 2.0L Turbo Diesel	
Renault Clio II 1.5 & 1.9 Diesel 2001-2006	9781898780748
Renault Clio II 1.2, 1.4, & 2.0 Litre Petrol 2001-2006	9781898780489
Renault Espace III & IV 1.9 Diesel from 1999-2005	9781907070068
Renault Kangoo, Nissan Kubistar 1.5 & 1.9L Diesel Models 2001-2006	9781898780960
Renault Kangoo, Nissan Kubistar 1.2 1.4 & 1.6L Petrol Models 2001-2006	9781898780175
Renault Trafic 1397cc, 1647cc, 1721cc, 1995cc and 2165cc Petrol Models	9781898780564
Renault Trafic II, Vauxhall Vivaro, Nissan Primastar	9781898780915
2.0 Litre 16V Petrol & 1.9 Litre dCi-DTI Diesel	
Suzuki 4WD, SJ410, SJ413, Jimny, Samurai & Vitara	9781898780694
1.0 JA & JL, 1.3 JX and 1.6 Litre 8 & 16 valve	
Volkswagen LT Types LT28, LT 35, LT 46 1997-2002	9781898780786
2.5 & 2.6L Diesel	
Volkswagen Saharan, Seat Alhambra 2001-2006	9781898780373
1.9 Litre TDi & 2.8 Litre 24 Valve VR6 Engine	

Made in United States
Orlando, FL
04 November 2026

90201279R00089